Introduction to Food Chemistry

Vassilis Kontogiorgos

Introduction to Food Chemistry

 Springer

Vassilis Kontogiorgos
Senior Lecturer in Food Chemistry
The University of Queensland, School of Agriculture and Food Sciences, St Lucia Campus
Brisbane, QLD, Australia

ISBN 978-3-030-85644-1   ISBN 978-3-030-85642-7   (eBook)
https://doi.org/10.1007/978-3-030-85642-7

This Springer imprint is published by the registered company Springer Nature Switzerland AG
The registered company address is: Gewerbestrasse 11, 6330 Cham, Switzerland

About this Book

The complexity of food chemistry makes it a challenging subject for students studying in a food science programme. Although there are excellent food chemistry books available in the market, they are either encyclopaedic or not pitched correctly to undergraduate food science students. These problems create difficulties for students to identify what is essential, how much they need to know or find links between the chemistry of food components and its relevance to applications.

The present textbook bridges this gap. It employs the latest pedagogical theories in textbook writing to present the subject to food science students and links chemistry with food processing, quality and shelf life.

From a student's perspective, the book has specific learning objectives for each chapter and is self-contained, so students do not need to search for essential information outside the textbook. The present book has didactic elements with information being conveyed with figures, colour-coded schemes and graphs, annotations on figures that link it to the text descriptions, and learning activities at the end of each chapter linked to the learning objectives.

Lecturers can also use this book to help them focus teaching preparation on key aspects of food chemistry relevant to both industry and modern research. The present textbook may also support lecturers with exam preparation, assignments, and other assessments or learning activities.

Some prerequisite knowledge is necessary to make the best use of the book. The reader must have knowledge of general chemistry and general organic chemistry to be able to follow the concepts presented in this textbook.

The book is organised into eight chapters covering structures, chemistry and functionality of all components found in food. Concepts from the physical chemistry of foods (e.g. glass transition and gelation) and food analysis are also introduced and explored in each chapter. At the end of each chapter, there are learning activities to help with the revision of essential concepts. Several exercises require access to the Internet, and occasionally more advanced concepts are introduced, and the learner can explore them at their own pace. Answers to some of the activities are provided in the Appendix. Additional food chemistry books and review articles on current food chemistry topics are presented in the bibliography. These resources have been used in one way or another in the preparation of the current textbook. None of the resources is older than 2000, and the majority of them are since 2010.

Now, let us get started with water and explore why it is so special for food stability.

The University of Queensland   Vassilis Kontogiorgos
School of Agriculture and Food Sciences
St Lucia Campus
Brisbane, Australia

Contents

1 Water 1
  1.1 Introduction 1
  1.2 Water and Ice Structure 2
  1.3 Hydrogen Bonding of Water 3
  1.4 Interactions of Water with Food Constituents 4
  1.5 Colligative Properties 6
  1.6 Water Activity 7
  1.7 Learning Activities 14
    1.7.1 Multiple Choice Questions 14
    1.7.2 Short Answer Questions – Further Reading 15
    1.7.3 Fill the Gaps 16

2 Carbohydrates 19
  2.1 Introduction 19
  2.2 Monosaccharide Structure 20
  2.3 Monosaccharide Reactions 26
  2.4 Disaccharides-Oligosaccharides 31
  2.5 Polysaccharides 32
    2.5.1 Starch 34
    2.5.2 Carrageenan 39
    2.5.3 Alginates 40
    2.5.4 Pectin 42
    2.5.5 Cellulose 43
    2.5.6 Galactomannans 43
    2.5.7 Gum Arabic 44
    2.5.8 Xanthan 45
    2.5.9 Chitin 46
    2.5.10 Dietary Fibre 46
  2.6 Learning Activities 47
    2.6.1 Multiple-Choice Questions – Monosaccharides 47
    2.6.2 Multiple-Choice Questions – Polysaccharides 49

2.6.3   Short Answer Questions – Further Reading.   50
2.6.4   Fill the Gaps.   52

3   Proteins-Enzymes   55
3.1   Introduction   55
3.2   Amino Acids   56
3.3   Proteins   61
3.3.1   Protein Classification.   61
3.3.2   Protein Structure   63
3.3.3   Changes in Protein Structure: Denaturation and
Hydrolysis   67
3.3.4   Functional Properties of Proteins.   72
3.4   Enzymes.   74
3.4.1   Mechanism of Enzymatic Reactions   74
3.4.2   Enzyme Kinetics   77
3.4.3   Enzyme Nomenclature and Classification   81
3.4.4   Food Enzymes   83
3.5   Learning Activities.   88
3.5.1   Multiple-Choice Questions – Proteins.   88
3.5.2   Multiple-Choice Questions – Enzymes   90
3.5.3   Short Answer Questions – Further Reading.   92
3.5.4   Fill the Gaps.   94

4   Lipids   97
4.1   Introduction   97
4.2   Fatty Acid Nomenclature and General Characteristics   98
4.3   Triacylglycerols (TAGs).   101
4.4   Lipid Oxidation   105
4.5   Fat Crystallisation   110
4.5.1   Crystal Formation   110
4.5.2   TAG Conformation and Polymorphism.   112
4.5.3   Other Properties of Fats   116
4.6   Learning Activities.   118
4.6.1   Multiple-Choice Questions   118
4.6.2   Short Answer Questions – Further Reading.   120
4.6.3   Fill the Gaps.   121

5   Browning Reactions   123
5.1   Introduction   123
5.2   Enzymatic Browning   124
5.2.1   Reactions and Substrates   125
5.2.2   Deactivation Strategies.   127
5.3   Non-enzymatic Browning   127
5.3.1   Caramelisation.   128
5.3.2   Ascorbic Acid Browning   130
5.3.3   Maillard Reaction.   130

    5.4   Learning Activities....................................... 136
         5.4.1   Multiple-Choice Questions 136
         5.4.2   Short Answer Questions– Further Reading 137
         5.4.3   Fill the Gaps.................................... 138

6   Vitamins-Minerals 141
    6.1   Introduction 141
    6.2   Fat-Soluble Vitamins 142
    6.3   Water-Soluble Vitamins 144
    6.4   Minerals.. 150
         6.4.1   Effect of Processing on Mineral Bioavailability 151
    6.5   Learning Activities....................................... 152
         6.5.1   Multiple-Choice Questions 152
         6.5.2   Short Answer Questions – Further Reading. 154
         6.5.3   Fill the Gaps.................................... 154

7   Colour Chemistry 157
    7.1   Introduction 157
    7.2   Interaction of Light with Food. 158
    7.3   Colour Chemistry... 160
         7.3.1   Myoglobin 161
         7.3.2   Chlorophyll 162
         7.3.3   Carotenoids 163
         7.3.4   Anthocyanins 165
         7.3.5   Bctalains....................................... 167
         7.3.6   Artificial Dyes 169
    7.4   Learning Activities....................................... 170
         7.4.1   Multiple-Choice Questions 170
         7.4.2   Short Answer Questions – Further Reading. 172
         7.4.3   Fill the Gaps.................................... 172

8   Flavour Chemistry 175
    8.1   Introduction 175
    8.2   Flavour Chemistry 176
    8.3   Flavour Sources 181
    8.4   Flavour Delivery 185
    8.5   Flavour Interactions and Stability 187
    8.6   Learning Activities....................................... 187
         8.6.1   Multiple-Choice Questions 187
         8.6.2   Short Answer Questions – Further Reading. 189
         8.6.3   Fill the Gaps.................................... 190

Appendices.. 191

Bibliography 195

Index... 201

Chapter 1
Water

Learning Objectives

After studying this chapter, you will be able to:

- Describe the structure of the water
- Describe the hydrogen bond and its importance for foods
- Describe the interactions of water with food constituents
- Discuss colligative properties and their importance in food processing
- Discuss the importance of water activity in food stability
- Assess chemical, physical, and microbiological food stability based on water activity values

1.1 Introduction

Water is an essential food constituent, and its content varies significantly across different products. For example, there are foods like fruits or meat products with a water content of ~95% or ~ 70%. In contrast, there are foods with a limited amount of water ranging between 10 and 20%. As several chemical and enzymatic reactions need water to occur, it serves as a solvent of water-soluble ingredients and a medium for diffusion. Water is also essential for the growth of beneficial (e.g., lactic acid bacteria) and spoilage (e.g., *Pseudomonas*) microorganisms. During food processing, it is crucial to calculate the correct amount of water in the formulation to achieve the specified product's consistency and optimum shelf life stability. For instance, minor deviations in water content in bakery products may result in quality losses and rejection of the final product. It is also of economic concern, as it adds weight to the product and increases shipping costs. It should be stressed that irrespectively of the processing, food always contains a certain amount of water that is not possible to remove entirely. For example, wafers or potato crisps that are particularly dry foods have a remaining water content of ~5%. As a result, controlling and predicting water behaviour is very important in food processing. The first step to control water in foods is to understand its structure, chemical, and physical properties. Especially important are the mechanisms with which it interacts with the various food constituents, as it is most frequently the underlying reason for product stability. This chapter presents all the essential information needed to explore the broad area of water functionality in foods.

1.2   Water and Ice Structure

When two hydrogen atoms form covalent bonds with an oxygen atom, they form water. Each hydrogen atom has a nucleus consisting of a proton surrounded by a negatively charged electron. The oxygen atom has eight electrons, and because of its higher *electronegativity*, oxygen attracts the shared electrons of the covalent bonds to a greater extent than hydrogen atoms. As a result, the oxygen atom acquires a *partial negative charge* (δ^-), while the hydrogen atoms acquire a *partial positive charge* (δ^+) (Fig. 1.1). *Polarity* is the separation of electric charges, making water a *polar molecule*. As a result of polarity, water molecules in the liquid phase interact with each other with short-lived (also known as *transient*) *hydrogen bonds* that result in unusual physical properties.

When temperature decreases below 0 °C water molecules arrange themselves in space in a hexagonal pattern creating the so-called hexagonal ice that is the form of ice present in foods (Fig. 1.2). Hexagonal ice has an open structure and lower density than liquid water and, as a result, ice floats (e.g., ice cubes in a cola drink). The expansion of water volume on freezing is a major cause of structural deterioration in frozen foods. As the ice expands, it essentially breaks down the structure, and in most cases, it cannot recover on thawing.

Fig. 1.1 Structure and partial charges of water. The angle between hydrogen atoms is about 104.5°. The letter δ indicates the partial charge

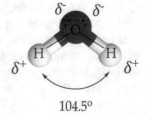

104.5°

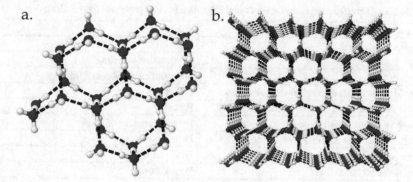

Fig. 1.2 Hexagonal ice: (**a**) Water molecules arrange in a hexagonal pattern at sub-zero temperatures, and (**b**) structure of an ice cube. Dashed lines represent hydrogen bonds between water molecules

Fig. 1.3 Hydrogen bonds of water. The dashed lines indicate hydrogen bonds between water molecules

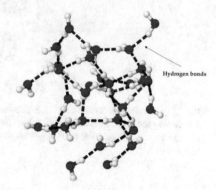

1.3 Hydrogen Bonding of Water

A hydrogen bond is a particular type of van der Waals attraction that results from attractive forces between a hydrogen atom covalently bonded to an electronegative atom of the first molecule (e.g., O, N or F) and an electronegative atom of a second molecule (Fig. 1.3). Hydrogen bonds are *intermolecular*, occurring between different molecules, or *intramolecular* occurring within the same molecule. For example, both types are found in proteins or polysaccharides, influencing the structure and the properties of these food components. Hydrogen bond (~20 kJ/mol) is weaker than covalent bond (~460 kJ/mol) but generally stronger than van der Waals attractions (~1.2 kJ/mol).

As a result of hydrogen bonding, water presents abnormal properties compared to molecules with similar molecular weight or structure, thus influencing food processing (Table 1.1). For example, with its unusually high boiling point (100 °C), it requires large amounts of energy to remove it during evaporation or dehydration. Additionally, volume increase on crystallisation damages food structure, which is important in frozen products.

Table 1.1 Abnormal properties of water and relevance to food processing operations

Unusual property	Process
High melting point	Freezing
High boiling point	Evaporation-dehydration
High critical point	Supercritical fluid extractions
Liquid water exists at very low temperatures	Freezing
Volume increases on crystallisation	Freezing
Large volume changes from liquid to gas	Evaporation-use of steam
High specific heat capacity (twice that of ice)	Heating-cooling
High heat of vaporization	Evaporation
High heat of sublimation	Freeze drying

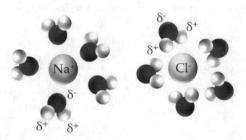

Fig. 1.4 Ionic interactions between water and NaCl. Water molecules separate Na^+ and Cl^- ions by breaking the ionic bond. Following ion detachment, water molecules surround the sodium and chloride atoms. Water surrounds the ions with the δ^+ side attracted to the Cl^-, and the δ^- side attracted to Na^+

1.4   Interactions of Water with Food Constituents

Water interacts with all substances found within food in different ways. The primary forms of interactions are *dipole-ion*, *dipole-dipole*, and *hydrophobic interactions*. The strength and extent of these interactions depend on the chemical nature of non-aqueous components, for example, salt concentration, pH, or temperature. A typical example of ionic interactions is between water and NaCl. Water arranges itself around ions to partially neutralise ionic charges separating positive and negative ions in aqueous solutions (Fig. 1.4).

The strength of water-ion interactions is usually greater than hydrogen bonding. As mentioned, water is a highly structured liquid because of the extensive network of hydrogen bonds. However, small ions with strong electric fields are *structure-forming*. They cause water molecules to rearrange themselves in the hydration shell around ions and exhibit a larger average density than pure water (e.g., Li^+, Na^+ or Ca^{2+}). In contrast, large ions with weak electric fields are *structure breaking* with the opposite effect (e.g., K^+, Cl^- or I^-). Although this may seem to be a mere scientific detail, it has enormous consequences for food fabrication and stability, as the ability

Fig. 1.5 Dipole-dipole interactions between (**a**) amino and carboxyl groups, (**b**) water and carboxyl group, and (**c**) water and amino group

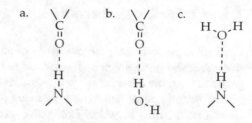

Fig. 1.6 Hydrophobic interactions between triglyceride molecules (in the middle) in water. Triglycerides interact closely with each other to exclude water molecules from their surroundings

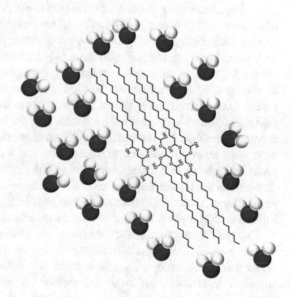

of ions to influence protein denaturation is linked to the way they interact with water (see Sect. 3.3.3, Hofmeister series). *Dipole-dipole interactions* refer to the interactions between water and non-ionic hydrophilic solutes found in food. For instance, molecules that carry hydroxyl (e.g., monosaccharides), amino (e.g., proteins), or carbonyl (e.g., flavours) groups may form hydrogen bonds with water (Fig. 1.5). Finally, hydrophobic interactions are the interactions between hydrophobic molecules in aqueous environments. Molecules with hydrophobic groups tend to exclude water from their surroundings and interact with each other (Fig. 1.6). Hydrophobic interactions are critical in proteins, lipids, and surfactants. They are also responsible for *gelation* events during food processing, as for example, in heat gelation of methylcellulose or egg-white proteins (e.g., egg boiling). Hydrophobic interactions are relatively stronger than van der Waals interactions and hydrogen bonds. *The strength of hydrophobic interactions increases at high temperatures, whereas the strength of hydrogen bonding increases with the lowering of temperature.*

1.5   Colligative Properties

Colligative properties are the properties of solutions that *depend on the number of molecules* in a given volume of water and *not* on their chemical nature. Colligative properties are the *lowering of vapour pressure, elevation of boiling point, depression of freezing point*, and *osmotic pressure* caused by the presence of small molecules that dissolve in water. Colligative properties may influence various food manufacturing processes (e.g., freezing or evaporation) or used for analytical purposes (e.g., examining for milk adulteration using cryoscopy). Sugars (e.g., glucose or sucrose) and salts (e.g., NaCl or $CaCl_2$) are frequently found in food formulations and can influence processing conditions. For example, during the concentration of a fruit juice, sugar concentration continuously increases with water evaporation. As a result, the boiling point of the remaining concentrated juice increases, requiring further heating to maintain evaporation. Another example is the freezing of ice-cream mix, where sugars present in formulation increase in concentration due to *freeze-concentration* in the unfrozen mix, and the melting point is depressed. As a result, to maintain ice cream structure formation, the temperature needs to be lowered even more. These two examples may be extrapolated to other foods that undergo high- or low-temperature processing. Ultimately, the presence of solutes determines the processing cost and the market price of food, as it controls energy expenditure. It starts to become evident that colligative properties and how they influence food processing need to be well-understood.

The underlying fundamental cause of the above events is the lowering of the vapour pressure of the solution *due to solute-water interactions*. Solutes immobilise water that cannot escape from the surface as vapour and, as a result, the higher the solute concentration, the lower the vapour pressure of a solution. *Raoult's law* describes the lowering of the vapour pressure. It states that lowering the vapour pressure is proportional to the number of particles in the solution. The vapour pressure of a solution is always lower than that of pure water. Freezing point depression ($\Delta T_f = K_f\,m$) and boiling point elevation ($\Delta T_b = K_b\,m$) are proportional to the *molality* (m) of the solution (i.e., amount of substance (moles) in a specified amount of mass of water (kg), $m = \text{moles}_{solute}/\text{mass}_{solvent}$) with K_f or K_b being the *cryoscopic* or *ebullioscopic* constants, respectively. Finally, osmosis is the diffusion of water through a semi-permeable membrane, from a solution of low solute concentration (hypotonic) to a solution of high solute concentration (hypertonic) (Fig. 1.7a). During this physical process water moves without energy input across a semi-permeable membrane (permeable to the water, but not to the solute), separating the two solutions of different concentrations (*forward osmosis*). This process is essential in cells in both animal and plant tissues and controls the movement of water between the cytoplasm and the surrounding layers of the cells. *Osmotic pressure* is the pressure we need to apply to the hypertonic solution to prevent osmosis (Fig. 1.7b). When the concentrations on either side are equal, then the solutions are called *isotonic*. Now, suppose the applied pressure exceeds the pressure required to prevent osmosis. In that case, the water flow is reversed, i.e., from the high

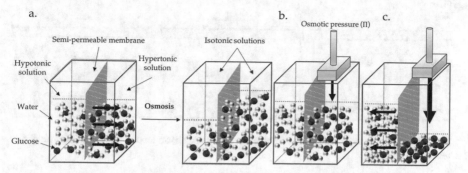

Fig. 1.7 (**a**) Osmosis, (**b**) osmotic pressure, and (**c**) reverse osmosis. See text for description

concentration solution to the low concentration solution (*reverse osmosis*) (Fig. 1.7c). Forward osmosis may be used in *osmotic dehydration* of some fruits. In contrast, reverse osmosis is a *membrane separation* technology used in water desalination (i.e., removal or NaCl from seawater). In addition, in foods that need syrup or brine (e.g., canned fruits or tuna), the concentration of sugars or salt needs adjustment to the correct levels to prevent osmotic dehydration of the product that may adversely affect quality.

1.6   Water Activity

Water content or *moisture content* is the quantity of water in food (g of H_2O/100 g of food). One may be eager to draw conclusions about long-term shelf life stability based on water content. However, various types of food with the same water content differ significantly in perishability. For example, marmalade and bread with a moisture content of ~33% w/w have quite different shelf lives, with bread losing its physical stability within hours and microbiological stability within a week. In contrast, marmalade may be stable for years after processing. As a result, water content is not a reliable predictor of microbial growth, chemical reactions, or structure stability of foods. The term water activity (a_w, dimensionless number, ranges from 0 to 1) accounts for the *intensity with which water associates with non-aqueous constituents* and is the ratio of the vapour pressure of water above food (p) over the vapour pressure of pure water (p_o), at the same temperature.

$$a_w(T) = p / p_o$$

In Fig. 1.8, water is in a sealed container that does not allow vapours to escape. Pure water at a specific temperature T reaches a *dynamic equilibrium* where the amount of water molecules that evaporate (liquid to gas) is the same as the number of water molecules that condense (gas to liquid) (Fig. 1.8a). Once equilibrium is established, vapours exert on the container pressure p_o and a_w is equal to 1. Inclusion

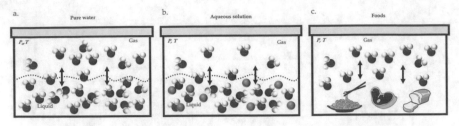

Fig. 1.8 Illustration of dynamic equilibrium in: (**a**) water, (**b**) aqueous solution, and (**c**) food. See text for description

of solutes, e.g., NaCl (green spheres, Fig. 1.8b), restricts evaporation of water and fewer molecules transfer to the gaseous state, creating pressure p. As a result, pressure p is always smaller than p_o and a_w is <1. In foods, the same principle exists, but the number of water molecules that escape to the gaseous state depends on the specific interactions of water with the food constituents (Fig. 1.8c). These interactions depend on the chemical composition of the product. Frequently, small changes in food formulation that are not perceivable by consumers allow a considerable increase in shelf life because of the changes in the strength of water interactions at the molecular level.

Using a_w as a predictor of microbiological, chemical, and physical stability, we can distinguish **three classes of foods**: *dry foods* with $a_w < {\sim}0.5$, *intermediate moisture foods* (IMF) with a_w between ~0.5 and ~ 0.8 and *high moisture foods* with $a_w > 0.8$. These three classes have different preservation requirements, and although the specific preservation details of each product vary depending on the composition, the following generalisations usually apply: dry foods are stable without refrigeration but need rehydration before consumption (e.g., pasta), IMF can be stored without refrigeration for prolonged periods but do not need rehydration before consumption (e.g., salami) whereas high moisture foods need refrigeration and no rehydration is necessary (e.g., yoghurt).

Moisture and Water Activity Determination

The *total moisture content of high moisture foods* is determined by drying (e.g., oven), where water is removed by heat and moisture is calculated from the weight loss. *Karl Fischer titration* is a method for water determination in low-moisture foods such as dried fruits and vegetables, confectionery products, chocolate, coffee, and other low-moisture foods high in fats, sugar, or protein. The method is based on the reaction between sulphur dioxide and iodine in the presence of water, imidazole (catalyst) and alcohol (solvent). Alcohol reacts with sulfur dioxide (SO_2) and base to form an intermediate alkylsulfite salt, which is then oxidised by iodine to an alkylsulfate salt. This oxidation reaction consumes water and iodine in a 1:1 ratio. Once all the water is consumed, the presence of excess iodine is detected titrimetrically.

The amount of water is calculated based on the concentration of iodine consumed in the titration.

ROH + **SO₂** + R´N⟶ [R´NH]SO₃R + **H₂O** + **I₂** + 2R´N⟶ 2[R´NH]I + [R´NH]SO₄R

[alcohol]   [base]   [alkylsulfite]   [iodine]   [hydroiodic acid salt] [alkylsulfate]

Water activity is measured with instruments designed for this purpose only. The principle of the method is based on measuring equilibrium relative humidity that surrounds the sample atmosphere at a constant temperature.

A plot of water content *vs* water activity is known as moisture sorption isotherm (MSI) and shows how water activity changes with addition (absorption) or removal (desorption) of water from **dry** foods at a specific temperature. Food engineers use this curve to design dehydration processes for foods, but we explore it from a different angle here. Foods where water uptake occurs readily, are called *hygroscopic*. In these foods, small changes in moisture result in considerable changes in water activity. Conversely, compounds that absorb large amounts of water with a minimal increase in water activity are called *anticaking agents*. Most foods lie between these two extremes, and their moisture sorption isotherm has a *sigmoid shape*. The shape of the moisture sorption isotherm is unique for each product. In this plot, *three zones* are distinguished corresponding to water with different physical properties (Fig. 1.9a).

In Zone 1 or *low moisture regime*, water has the least molecular mobility (i.e., the ability of water molecules to move and diffuse) and corresponds to a dry product with the highest shelf life stability. In this region, water molecules are in direct contact with food components and interact strongly with each other. As a result, they form a monomolecular layer with entirely different properties than bulk water (Fig. 1.10). This water type (Zone 1) is also known as *bound water*, but this term should be generally avoided, as it is thermodynamically incorrect and may be misleading. Dehydration cannot remove Zone 1 water from food and, additionally, it does not freeze under normal industrial freezing processes (~ −40 °C). Additionally, it cannot act as a solvent, support chemical reactions, or sustain microbial growth. Even the driest of foods, for example, pasta, crisps, or food powders, always have a residual water content in the range of ~5–10% w/w. Frozen foods stored at −18 °C have substantial amounts of water in the liquid state. In Zone 2, additional layers of water accumulate on top of the first layer forming *multiple layers* of water having *intermediate molecular mobility* (Fig. 1.10). This water is not as hard to remove or freeze as Zone 1 water. Intermediate moisture foods such as dried fruits, hard cheeses or cured meat products are typical foods containing multilayer water. Finally, water in Zone 3 has the *highest molecular mobility* and its physical properties do not differ substantially from the properties of bulk water (Fig. 1.10). As a

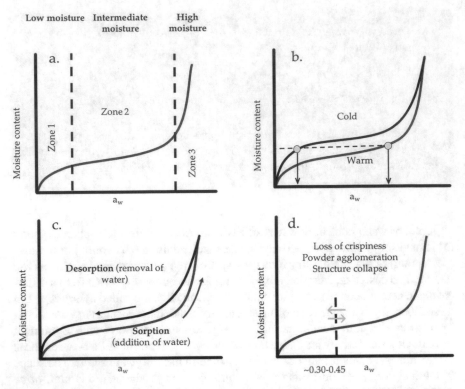

Fig. 1.9 Moisture sorption isotherms of foods (**a**) the three zones of water with different molecular mobility, (**b**) influence of temperature on MSI. Storage at high temperatures increases a_w, (**c**) influence of addition or removal of water on MSI. The lack of curve overlap is termed hysteresis, and (**d**) texture changes above a critical a_w. Poor control of critical a_w leads to texture problems

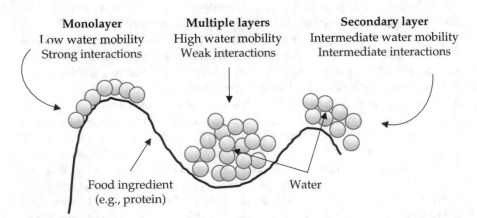

Fig. 1.10 Water layering corresponding to different zones of an MSI (Fig 1.9a). Zone I – monolayer, zone II -secondary layer, and zone 3 – multiple layer

Table 1.2 Approximate critical a_w values for the onset of chemical, physical or microbiological changes

Critical a_w	Event
~0.2–0.3	Lipid oxidation rate increases
~0.2–0.3	Reactions requiring water do not occur
~0.35–0.4	Powder agglomeration
~0.4–0.5	Loss of crispiness when moisture is absorbed
~0.5	Onset of hardening when moisture is lost
~0.6	Onset of microbial growth
~0.85	Onset of pathogen growth

result, it can act as a solvent, influence the rate of chemical reactions and microbial growth, and cause changes in the overall structural stability and quality of foods.

It should be emphasised that a_w *depends on temperature*. An increase in temperature increases a_w for the same moisture content due to higher water mobility resulting in loss of stability (Fig. 1.9b). Consequently, the *temperature should always be maintained during storage* of dry foods to avoid chemical, physical and potentially microbiological changes in the product. Moisture sorption isotherms prepared by the addition of water (rehydration) to a dehydrated food are different from those prepared by water removal (dehydration). This phenomenon, called *hysteresis*, occurs due to permanent structural changes in food during drying and may affect the texture and other sensory properties of products on rehydration and consumption (Fig. 1.9c). When a_w exceeds a *critical* value, food usually loses its structural stability. For example, crispiness in dry bakery products or the ability of powders to flow is compromised when a_w exceeds a specific value, usually ~0.30 (Fig. 1.9d, Table 1.2).

Glass Transition

Solids can be either crystalline or amorphous. The crystal structure is highly organised, and molecules have specific locations in space. In amorphous materials, molecules are disordered, meaning they do not have an exact position in space, i.e., they are randomly positioned. The simplest food example is table sugar (sucrose). Table sugar exists as both crystalline and amorphous solid. As bought from the market, table sugar is crystalline, and the sugar molecules have specific positions within each sugar grain. If we add very little water (or melt it), it dissolves and forms a very think syrup. If we cool this thick syrup, it forms a hard candy. Even though the hard candy consists of the same sugar molecules, it will be amorphous and not crystalline, as the sucrose molecules are now randomly positioned within the candy.

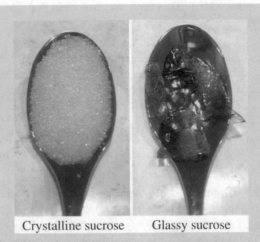

Crystalline sucrose   Glassy sucrose

The transformation of the sugar syrup to amorphous hard candy is known as *glass transition*. Conversely, the hard candy can also be heated and turn back into a thick syrup; this is also a glass transition. Amorphous materials are also known as *glassy*. Glass transition is one of the most important physico-chemical characteristics of non-crystalline, amorphous food solids as it controls the functionality and stability of the food product.

In addition to simple sugars, a lot of major food ingredients are amorphous under certain conditions, including proteins and polysaccharides. Glass transition is very important in foods with low water content, for example, breakfast cereals, frozen foods, bakery and confectionery products, food powders or extruded foods. Glass transition is not important in foods with high water content. Tiny changes in water content (e.g., through absorption from the atmosphere) result in significant losses of product quality even though it may not be a health hazard. Examples are softening of wafers or sticking of powder particles with moisture absorption. When wafers soften, they have undergone a glass transition! This softening is termed *plasticisation,* and water is the most common food *plasticiser* responsible for undesirable, in most cases, softening. Water can be thought of as a "lubricant" that lubricates molecules and facilitates their movement. Lipids do not undergo glass transition.

Moisture migration in *multicomponent foods* influences physical stability, as migration from a high (product A in Fig. 1.11) to a low (product B in Fig. 1.11) moisture region occurs continuously until a thermodynamic equilibrium with the surrounding environment is established (Fig. 1.11). Techniques to slow this process or prevent it altogether from occurring include the addition of edible layers between food components to separate the regions with different moisture contents physically (e.g., a chocolate layer on the inside of an ice cream cone to reduce the moisture

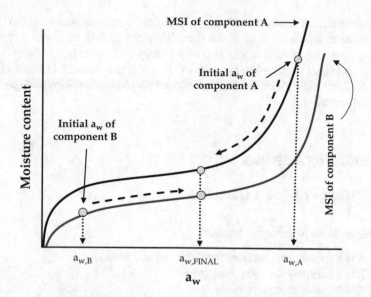

Fig. 1.11 Moisture sorption isotherms of multicomponent foods. See text for explanation

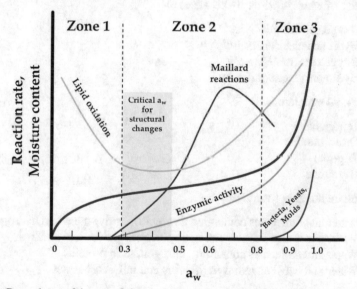

Fig. 1.12 Dependency of the rate of chemical, physical and microbiological changes on a_w

migration and maintain crispness during storage) or matching the a_w of food components (e.g., raisins and freeze-dried fruits in breakfast cereals).

Figure 1.12 shows a convenient plot with the reaction rates of important chemical, physical and microbiological changes as a function of a_w. From this plot, some generalisations may be drawn: (i) the rates of microbial growth, enzymatic activity, and hydrolytic reactions decrease with a_w, (ii) Maillard reactions peak at $a_w \sim 0.7$, (iii) lipid oxidation increases at $a_w < 0.3$ (i.e., in dehydrated foods), and (iv) loss of crispiness, powder stickiness and structure collapse occur at a_w between 0.3–0.5. Although the exact relationship between the reaction rates and the changes outlined in Fig. 1.12 are highly specific for each food, this plot presents helpful guidelines worth memorisation.

1.7   Learning Activities

1.7.1   Multiple Choice Questions

1. Water in foods is important because:

   (a) It is a solvent and medium for diffusion and reactions
   (b) It is a substrate for microbial growth
   (c) It influences storage stability
   (d) All of the above

2. Polarity of water is primarily because of:

   (a) Oxygen electropositivity
   (b) Hydrogen electronegativity
   (c) Oxygen electronegativity
   (d) Hydrogen electropositivity

3. Ice crystal structure is:

   (a) Tetragonal
   (b) Orthorhombic
   (c) Trigonal
   (d) Hexagonal

4. Hydrogen bonding forms between

   (a) Water and molecules containing an electropositive atom and hydrogen atom
   (b) Water and molecules containing an electronegative atom and hydrogen atom
   (c) Water and molecules containing nitrogen, as in proteins
   (d) Water and between atoms in proteins and polysaccharides

5. Hydrogen bond is

   (a) Always an attractive electrostatic force
   (b) Always a repulsive electrostatic force
   (c) Attractive between in water molecules but repulsive in proteins
   (d) Repulsive between in water molecules but attractive in proteins

6. Arrange the following bonds in order of increasing bond strength:

   (a) (stronger) Covalent > van der Waals > Hydrogen (weaker)
   (b) (stronger) Covalent > Hydrogen > van der Waals (weaker)
   (c) (stronger) Hydrogen > van der Waals > Covalent (weaker)
   (d) (stronger) van der Waals > Hydrogen > Covalent (weaker)

7. Colligative properties of solutions are:

   (a) elevation of boiling point, elevation of freezing point and depression of
       water activity
   (b) depression of boiling point, elevation of freezing point and osmosis
   (c) elevation of boiling point, depression of freezing point and osmosis
   (d) depression of boiling point, elevation of freezing point and elevation of
       water activity

8. The importance of colligative properties is that:

   (a) the presence of sugars and salt controls the processing properties of foods
   (b) controlling of boiling point extends shelf life
   (c) minimise spoilage from bacteria
   (d) control water activity, especially in high moisture foods

9. Water activity and water content values of food are:

   (a) always the same
   (b) always different
   (c) cannot be compared as they are different properties
   (d) always different ONLY in dry foods

10. Drying of foods creates products with water activity:

   (a) < 0.3
   (b) $0.3 < a_w < 0.5$
   (c) $0.5 < a_w < 0.9$
   (d) >0.9

1.7.2 Short Answer Questions – Further Reading

1. Discuss the abnormal physical properties of water because of hydrogen bonding
   and its relevance to food processing.

2. Discuss the interactions that occur between water and food constituents. Give
   one example of each interaction and its importance in food processing.

3. Describe osmosis and define osmotic pressure. Use a drawing to support
   your answer.

4. Describe the dynamic equilibrium between vapour and liquid water above food
   in a closed container at a constant temperature.

5. Draw a moisture sorption isotherm and identify the three zones of water. Discuss the type of water and the strength of interactions between water and food constituents in each zone.

6. An apple pie has three components: pastry, jam, and apple pieces. Initial formulation resulted in a soggy product with a short shelf life because of the differences in a_w of the components. Discuss how moisture migration occurs in multicomponent foods with different water activities. What steps could you take to minimise moisture migration and extend the shelf life of this product?

7. **Online activity, Advanced** Search online for the Clausius-Clapeyron relation and discuss its relevance to water activity.

8. **Advanced** Describe the stability of skimmed milk powder using the glass transition concept: a) as humidity increases at a constant temperature, and 2) as temperature increases at constant humidity.

9. Discuss the applications of glass transition in food processing.

10. **Online activity** Find online the terms "first-order transition" and "second-order transition". Identify the differences between them and give two examples for each.

11. **Advanced, Further reading**: Find the following review article "Roos, Y. H. (2010). Glass Transition Temperature and Its Relevance in Food Processing. Annual Review of Food Science and Technology, 1(1), 469-496."
    - What are α- and β-relaxations?
    - What are the main theories of glass transition?
    - What experimental techniques can we use to measure glass transition?
    - What is a "state diagram"?
    - How are glasses formed during food processing (Fig. 7, page 481-onwards)?

1.7.3 Fill the Gaps

1. The separation of electric charges is called _____ making water a_____ molecule.
2. Because of _____ _____ water presents abnormal properties that influence food processing.
3. _____ interactions are the interactions between hydrophobic molecules in aqueous environments.
4. _____ is the diffusion of water through a semi-permeable membrane
5. Water activity accounts for the _____with which water associates with various non-aqueous constituents.
6. In intermediate moisture foods, a_w is between _____ and ____

7. In Zone 1 water has the least _____ _____ and corresponds to a product with the _____ stability.
8. Increase in temperature _____ a_w for the same moisture content.
9. The transformation of the amorphous hard candy to thick syrup on heating is known as _____ _____.
10. The softening of wafers is termed _____, and water is the most common food _____.

Chapter 2
Carbohydrates

Learning Objectives

After studying this chapter, you will be able to:

- Describe the structure of monosaccharides
- Discuss monosaccharide reactions
- Identify the different types of glycosidic bonds in polysaccharides
- Explain the changes that occur in starch during processing
- Identify food polysaccharides based on their chemical structure
- Discuss the functional properties of polysaccharides

2.1 Introduction

Carbohydrates are structural components of living organisms, e.g., cellulose in plants or chitin in arthropods and a significant energy source through oxidation for both plants and animals. In foods, they may be used as sweeteners, emulsion stabilisers, gelation agents, or to retain water in food formulations. Chemically, carbohydrates are polyhydroxy aldehydes or polyhydroxy ketones and are classified into three groups. Monosaccharides are the simplest form of carbohydrates and the building blocks of the rest of carbohydrate-based molecules. Linking monosaccharides creates larger molecules such as disaccharides consisting of two monosaccharides or oligosaccharides comprised of between 3 and 20 monosaccharides. Finally, polysaccharides consist of more than 20 monosaccharide units, most frequently more than 200 and have distinct properties from the other two classes. This chapter introduces the chemical structure of carbohydrates relevant to foods and how their interactions with other molecules may influence product stability and quality.

2.2 Monosaccharide Structure

Monosaccharides are either *aldoses* (e.g., glucose) or *ketoses* (e.g., fructose) when they carry an aldehyde or ketone functional group, respectively. Fructose is the only ketose relevant from a food chemistry perspective and discussed in this textbook. Monosaccharides have several carbon atoms, usually between three and six, but sometimes even more. Those relevant to food processing have six (e.g., glucose or fructose) or five (e.g., arabinose or xylose) carbon atoms. Monosaccharides with fewer than five carbon atoms or more than six are essentially outside the scope of food chemistry. In the context of food science, *the six most essential monosaccharides* are glucose, mannose, galactose, arabinose, xylose, and fructose. A few more relevant for specialised applications (e.g., erythrose or ribose) or an in-depth description of polysaccharide structures (e.g., gulose or rhamnose) will be mentioned when necessary. It is essential to know that *the numbering of carbon atoms starts from the carbonyl group* (Fig. 2.1). Understanding chemical reactions, polysaccharide linkage, and other carbohydrate properties is problematic without correct carbon atom numbering.

Because of the intricacy of their structure, carbohydrates have several isomeric forms (Fig. 2.2). Specifically, monosaccharides show a form of isomerism termed *stereoisomerism*, where atoms connect in the same order but differ in spatial arrangement (i.e., 3D arrangement). Stereoisomerism is divided into *enantiomerism* and *diastereoisomerism*. In enantiomerism, the structures are *non-superimposable mirror images* of each other and the isomers are termed *enantiomers*. In *diastereoisomerism*, the isomers are **not** mirror images of each other and are termed *diastereoisomers*. Diastereoisomers are further divided into *epimers* and *anomers*. Epimers differ in the orientation of a -OH group at *only one* stereocenter other than the anomeric carbon. Anomers differ at a new chiral carbon atom formed on ring formation (see below for a detailed explanation of the terms).

Carbohydrates show such complex isomeric forms because of the presence of multiple *chiral carbons* in their structure. A carbon atom of a molecule with four different groups attached to it is called chiral carbon and indicated as C*. In the simplest case, a molecule with four different substituent groups has two isomers where one structure is the reflection of the other (Fig. 2.3). These structures are

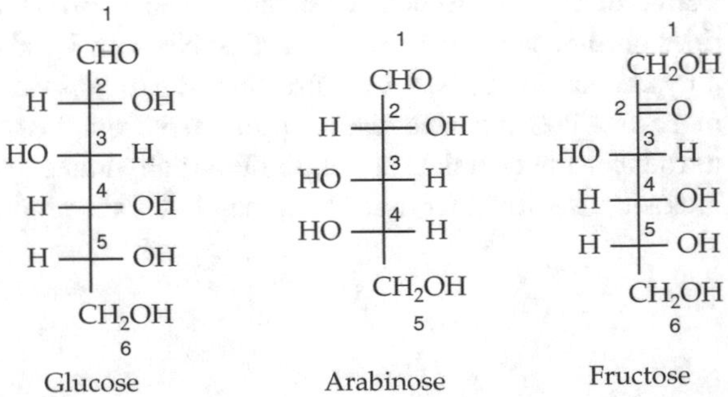

Fig. 2.1 Glucose, arabinose (aldoses) and fructose (ketose). The numbering of carbon atoms starts from the carbon atom with the carbonyl group

Glucose   Arabinose   Fructose

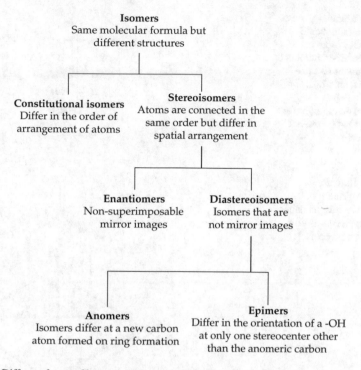

Isomers
Same molecular formula but
different structures

Constitutional isomers
Differ in the order of
arrangement of atoms

Stereoisomers
Atoms are connected in the
same order but differ in
spatial arrangement

Enantiomers
Non-superimposable
mirror images

Diastereoisomers
Isomers that are
not mirror images

Anomers
Isomers differ at a new carbon
atom formed on ring formation

Epimers
Differ in the orientation of a -OH
at only one stereocenter other
than the anomeric carbon

Fig. 2.2 Different forms of isomerism encountered in monosaccharides

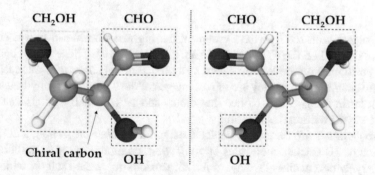

Fig. 2.3 Chiral carbon and enantiomers of glyceraldehyde. The structures are non-superimposable mirror images of each other

called *enantiomers* and are *non-superimposable mirror images* of each other. Non-superimposable means that when one structure is placed over the other, they do not overlap. Sugars contain one or more chiral carbon atoms, so they exhibit 2^n different arrangements of atoms, with n being the number of chiral carbons. For example, glucose with four chiral carbons has $2^4 = 16$ different isomers.

Fig. 2.4 D- and L-glucose.
The highest numbered
chiral carbon atom in
glucose is carbon number 5
(C-5). Carbons C-1 and
C-6 are not chiral. At C-5,
the –OH group is
positioned on the right for
D-glucose and on the left
for L-glucose

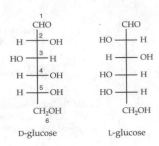

Fig. 2.5 Drawing the
Fischer projection of
monosaccharides on paper.
The same process is also
used for amino acids

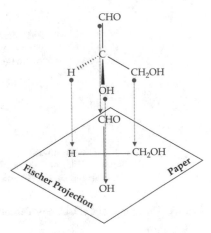

All sugars with the hydroxyl group on the **highest numbered chiral carbon atom** positioned on the right-hand side are called *D-sugars*. Those with a hydroxyl group positioned on the left are called *L-sugars* (Fig. 2.4). D- and L- monosaccharides are enantiomers, as they are mirror images of each other. Carbohydrates naturally occur in the D- form. Notice that the D- and L- are small capitals that is the correct way of writing this notation.

Carbohydrates are frequently depicted using the *Fischer projection*, a 2D representation of 3D organic molecules by *projection*, commonly used in carbohydrate chemistry. Monosaccharides have three dimensions in space (length, width, and height). However, we need to draw monosaccharides on paper, which has only two dimensions (length and width). In that case, to obtain the Fischer projection, the substituent groups of carbons are mirrored onto a 2D plane (paper) using straight lines (Fig. 2.5). The monosaccharide 2D image formed by joining the substituent groups on the plane of the paper is termed *projection*. Fischer projection is also used to represent amino acids (see Chap. 3). This structure is obtained by placing **the carbonyl group on top** and the rest of the carbon chain at the bottom (Fig. 2.4).

Monosaccharides carry aldehyde and alcohol groups in their structure, and as a result, they may react *intramolecularly* to form *hemiacetals*. Further reaction of hemiacetals with alcohols results in *acetal* (Fig. 2.6). This reaction forms a *cyclic*

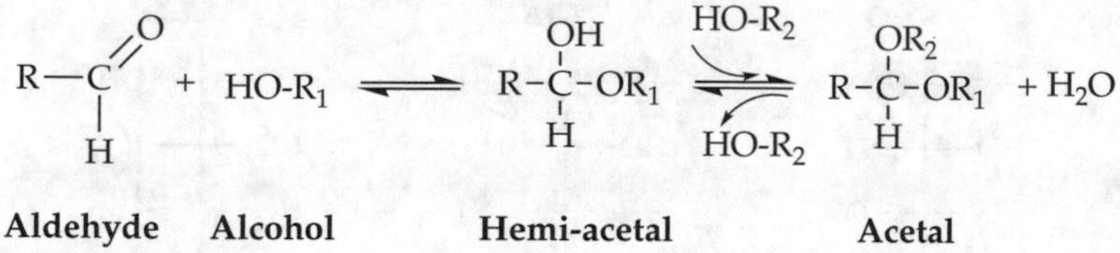

Fig. 2.6 Hemiacetal and acetal formation. The reaction of an aldehyde with alcohol results in hemiacetal formation. The reaction of hemiacetals with alcohols results in acetal formation

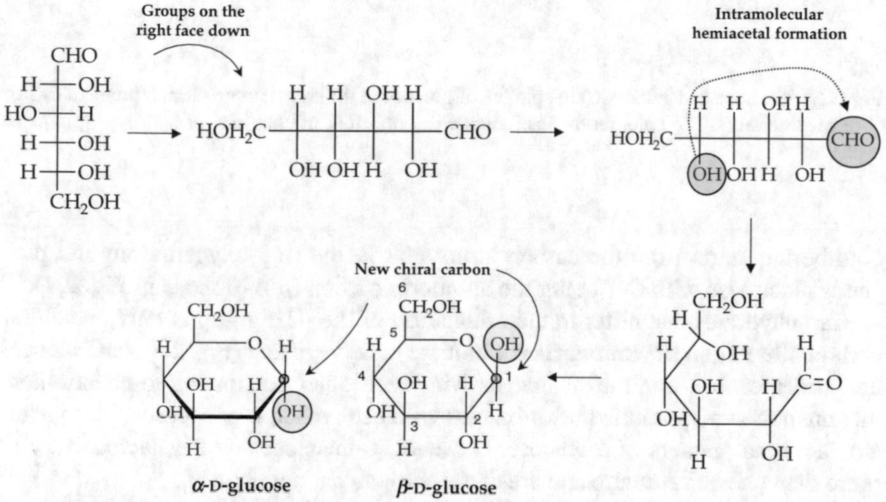

Fig. 2.7 Formation of Haworth projection and anomers through intramolecular hemiacetal formation. Starting from Fischer projection, the structure is rotated 90° clockwise. The OH group of C-5 reacts intramolecularly with the aldehyde group (highlighted in blue). Cyclisation produces a new chiral centre (green circle), forming α- and β-anomers. Positioning of hydroxyl groups (highlighted in yellow) is the only difference between the structures

hemiacetal (for aldehydes, e.g., glucose) or *hemiketal* (in ketones, e.g., fructose) and a *new chiral carbon*. Monosaccharides can form 6- or 5-membered rings called *pyranose rings* (e.g., glucose) or *furanose rings* (e.g., fructose). The *new* chiral carbon is called *anomeric carbon*, which results in two stereoisomers (alpha (α-) and beta (β-)) that differ only in their configuration at the anomeric carbon. A *Haworth projection* is another type of projection used to represent the cyclic structures of monosaccharides. In this projection, carbon and hydrogen atoms sometimes are not depicted, and the thicker bonds indicate that the atoms are closer to the observer (a-D-glucose in Figs. 2.7 and 2.9). In the Haworth projection, the *alpha-* (α-) anomer of D-monosaccharides have the -OH group *below the plane of the ring*, whereas *beta-* (β-) anomers have the -OH group *above the plane of the ring* (Fig. 2.7). It is important to note that for the L- sugars the opposite is true.

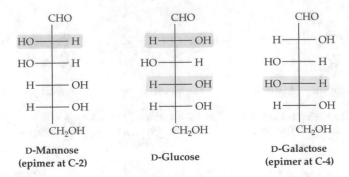

Fig. 2.8 Mannose and galactose are epimers of glucose at different carbon atom (epimer at C-2 or C-4, respectively). The coloured boxes indicate the different orientations of hydroxyl groups in each case

Numbering starts from the carbon atom next to the ring oxygen atom and proceeds clockwise with C-1 being the anomeric carbon (β-D-glucose in Fig. 2.7).

Carbohydrates that differ in the orientation of the -OH group at only one chiral carbon **other than the anomeric carbon** are called *epimers* (Fig. 2.8). Epimers are non-superposable, non-mirror images with each other contrasting to enantiomers that are non-superposable, mirror images with each other. D-Mannose and D-galactose are both epimers of D-glucose. However, D-mannose and D-galactose vary at more than one chiral centre and are diastereomers but not epimers.

The Haworth projection is not a realistic representation of monosaccharide structure in space because –OH groups and –H atoms interact and repel each other to minimise interactions between them. The 3D arrangement of atoms in space is termed *conformation* and molecules may convert from one form to another *by rotation* about individual single bonds. The various structures that result from the different arrangements of the substituent groups in space are termed *conformers*. Because of geometrical restrictions collectively termed *steric hindrances* (e.g., bond length, size of substituent groups etc.), the available number of the 3D shapes is limited. The two most common conformers of pyranoses are the so-called *chair* and *boat* conformers (Fig. 2.9). The most stable is the chair, whereas the boat conformer is seen only in derivatives with very bulky substituents.

Chair conformation is a common way of depicting carbohydrates, and we need to discuss it a bit more. There are two positions for the substituent groups for each carbon atom in these structures: the *axial* and the *equatorial* (Fig. 2.10). Axial bonds are parallel, whereas equatorial bonds are perpendicular to the axis of the ring. Each carbon has one axial and one equatorial bond. Bulky groups are in equatorial positions, whereas hydrogen atoms are in axial positions, as this positioning is less crowded. Two types of chair conformations may be formed depending on the positioning of C-1 and C-4. When C-4 is raised above the ring plane with C-1 below the

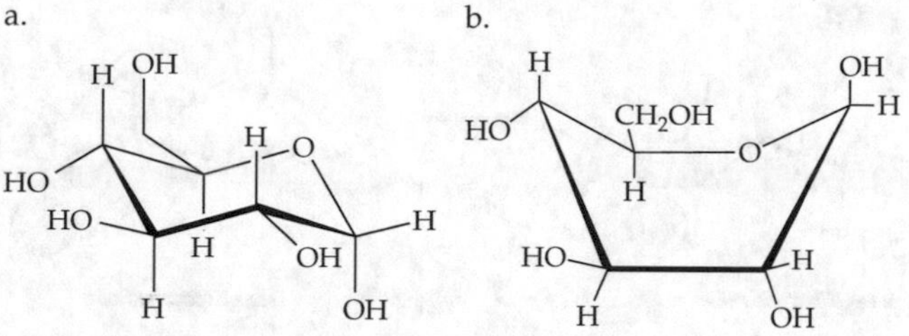

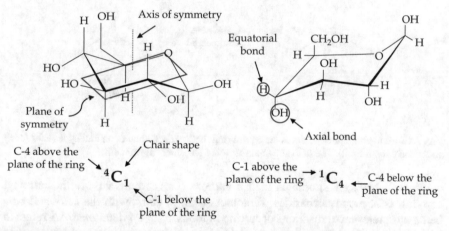

Fig. 2.9 Conformation of pyranoses, (**a**) chair, and (**b**) boat. The thicker bonds indicate that the atoms are closer to the observer

Fig. 2.10 Chair conformations of β-D-glucopyranose. Depending on the positioning of C-1 and C-4 we may have $^4C_1$ or $^1C_4$ conformations. See text for explanation

ring plane then this conformation is designated as $^4C_1$. When the opposite is true, then it is designated as $^1C_4$ (Fig. 2.10).

When monosaccharides dissolve in water, it results in an equilibrium mixture of open chain, five-, and six-membered rings where the configuration of the anomeric carbons may be either α- or β-. The anomers interconvert (i.e., $\alpha \rightleftharpoons \beta$) through their open form until an equilibrium is reached. This process is termed *mutarotation* because of the continuous change in the optical rotation of linearly polarised light (Fig. 2.11). For example, the equilibrium mixture of β-D-glucopyranose is ~64% of β-D-glucopyranose and ~ 36% of α-D-glucopyranose. In practical terms, it means that when we dissolve monosaccharides in water, we never have only one structure in the solution but a mixture of various forms depending on the monosaccharide and the surrounding environment (e.g., pH, temperature etc.).

Tautomers are isomers of compounds that interconvert from one form to another after intramolecular proton transfer from carbon to oxygen. This isomerisation reaction termed *tautomerisation* is catalysed by acid or base. It requires carbonyl

α-D-glucopyranose   β-D-glucopyranose

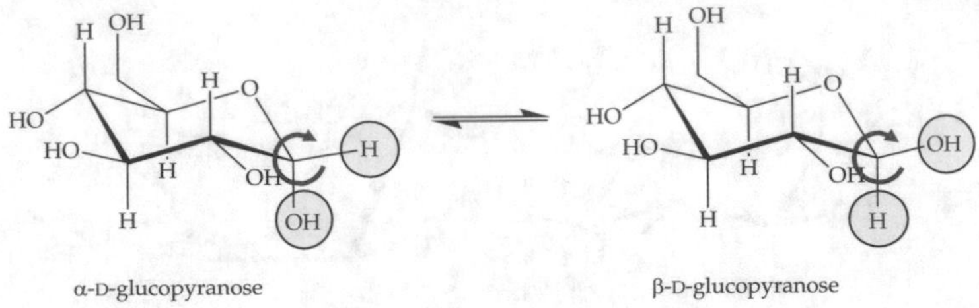

Fig. 2.11 When pure anomers dissolve in water, they undergo mutarotation interconverting from one anomer to another until an equilibrium is reached

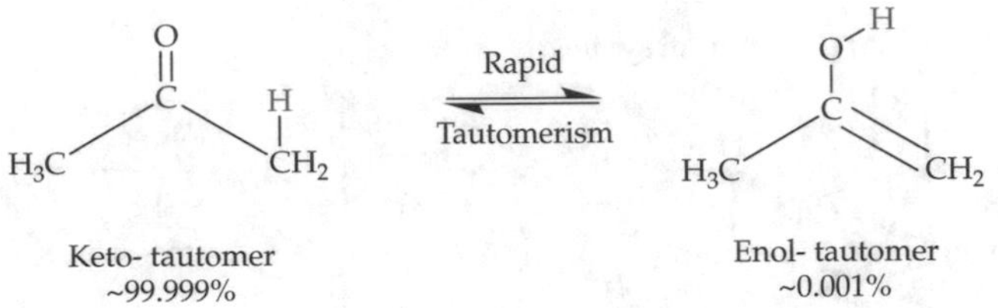

Keto- tautomer
~99.999%

Enol- tautomer
~0.001%

Fig. 2.12 Keto-enol tautomerism is the equilibrium between a ketone or an aldehyde (keto form) and alcohol (enol form). The keto and enol forms are tautomers of each other

compounds having H-atoms on their α-carbons (i.e., carbons next to the carbonyl carbon), as in monosaccharides. Tautomers exist in pairs, with the *keto-enol pair* being most relevant to reactions occurring in foods. *Keto-enol tautomerism* refers to interconversion between a *keto* form (ketone or aldehyde) and an *enol* (unsaturated alcohol) (Fig. 2.12). The interconversion is very fast with the keto form being the most stable. In the base-catalysed keto-enol tautomerism of D-glucose, the interconversions involve the movement of a proton and the shifting of bonding electrons. Consequently, when D-glucose is dissolved in water in alkaline conditions the solution also contains D-mannose and D-fructose! Keto-enol tautomerism occurs in the non-enzymatic browning of foods, as it is involved in Maillard reactions.

2.3 Monosaccharide Reactions

Reactions of monosaccharides are crucial in food processing and influence the quality and shelf life of products. Monosaccharide reactions involve both hydroxyl and carbonyl groups. Alcohol (hydroxyl group) may react to form *esters* and *ethers* or can be oxidised to form *acids*. Ketones can be reduced to *secondary alcohols*. In *esterification*, the reaction of alcohol and acid, hydroxyl groups of monosaccharides may be esterified. The reaction of alcohol with halides (compounds containing F,

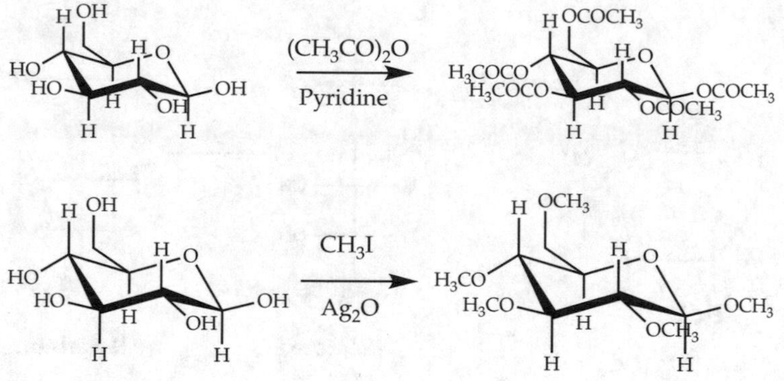

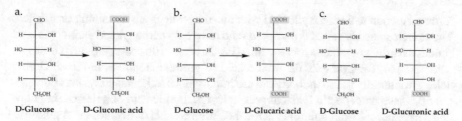

Fig. 2.13   Esterification (top) and etherification (bottom) reactions of carbohydrates

Fig. 2.14   Oxidation of glucose to (**a**) aldonic acid, (**b**) aldaric acid, and (**c**) uronic acid

Cl, Br, or I) results in *ether* formation (Fig. 2.13). These two reactions are essential in the commercial production of modified starches or cellulose derivatives and the manufacturing of emulsifiers (e.g., sorbitan esters) used in various food applications.

The aldehyde group can be *oxidised* to the corresponding carboxylic acid to form *aldonic acid*. When an aldose converts to aldonic acid, then the suffix "-onic" is used. For example, the acid of glucose is *gluconic acid* (Fig. 2.14a). After isomerisation, ketoses can also be oxidised to form glucose or mannose *via* an intermediate, *enediol*, through keto-enol tautomerism described earlier. Aldonic acids may also cyclise through an interesterification reaction between the carboxyl group and the –OH at C-5 to form *lactones*. The lactone of glucose, termed *glucono delta-lactone*, (GdL) is a common food additive. GdL hydrolyses slowly to gluconic acid, ensuring a progressive decrease of pH that resembles the acidification produced in bacterial fermentations. GdL is used in many foods, including meat, dairy, soy, or confectionery products. Under stronger oxidising conditions, both the aldehyde and the terminal -OH groups oxidise to carboxylic acids to form *aldaric acids*. In this case, the suffix "-aric" is used, for example, the acid of glucose is *glucaric acid* (Fig. 2.14b). The final scenario is when oxidation occurs in only the -OH of the

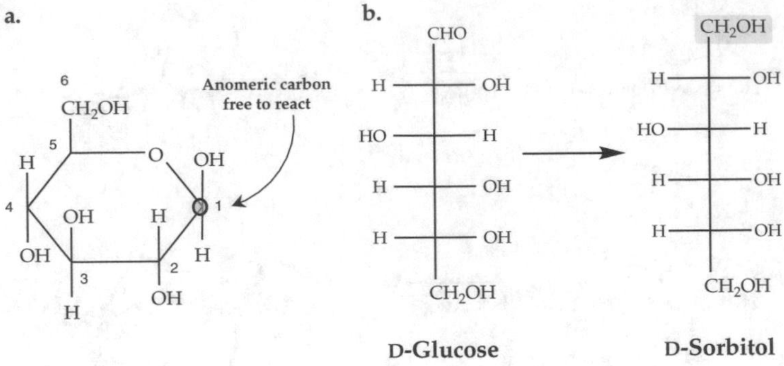

a.

b.

D-Glucose   D-Sorbitol

Fig. 2.15 (**a**) The anomeric carbon C-1 is free to react in monosaccharides. All monosaccharides are reducing sugars, and (**b**) reduction of monosaccharides forms alditols

terminal carbon to a carboxylic acid with the aldehyde group remaining unaffected. These molecules are called *uronic acids* and have both carbonyl and carboxylic acid functional groups available for reactions (Fig. 2.14c). In this case, the suffix "-uronic acid" is used, for example, the uronic acid of glucose is *glucuronic acid*. Uronic acids are naturally found in foods and are important constituents in various polysaccharides such as in pectin (galacturonic acid, i.e., uronic acid of galactose) or alginates (mannuronic and guluronic acids, i.e., uronic acids of mannose and gulose).

When monosaccharides oxidise (e.g., glucose, fructose, or galactose), another compound is reduced. Because of this property, monosaccharides are also known as *reducing sugars* (Fig. 2.15a). Some disaccharides (e.g., lactose or maltose), oligosaccharides and polysaccharides that have the anomeric carbon free to react (free aldehyde or ketone group) are also reducing. In contrast, sucrose is not reducing sugar because the anomeric carbon is involved in the glycosidic bond between glucose and fructose. Oxidation is also used for reducing sugar determination using the *Fehling's or Benedict's reagents*.

In *reduction*, carbonyl groups (aldehydes or ketones) are reduced to form *sugar alcohols* termed *alditols* using the suffix "-itol" (Fig. 2.15b). For example, sorbitol (from glucose, also known as glucitol), mannitol (from mannose) or xylitol (from xylose) are commonly used as non-calorific sweeteners in the confectionery industry to create sugar-free products (e.g., sugar-free chewing gums).

A reaction of central importance in carbohydrate chemistry is the *glycosidic bond* formation that joins two monosaccharides or very frequently a monosaccharide to other molecules. This bond is essential in the formation of di-, oligo-, and polysaccharides. In glycosidic bond formation, the hemiacetal in the anomeric carbon (C-1) reacts with alcohol to form an acetal called *glycoside*. If the -OH does not belong to another monosaccharide then the group is termed *aglycone* (Fig. 2.16). Glycosides are most frequently linked *via* an oxygen atom (*O*-glycoside), but they

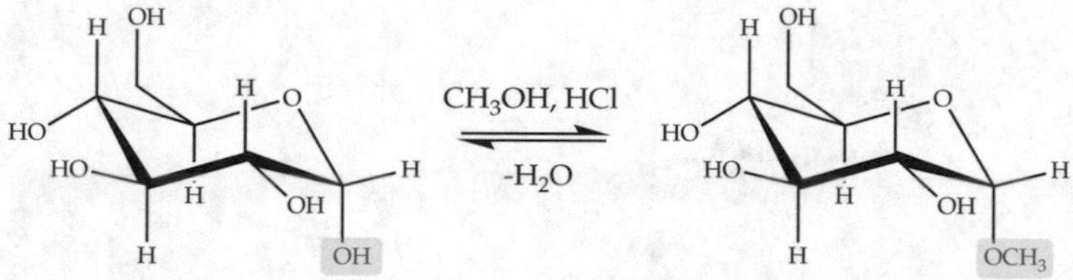

Fig. 2.16 Glucose reacts with methanol to form the glycoside methyl-glucose. In this case, the methyl group is the aglycone

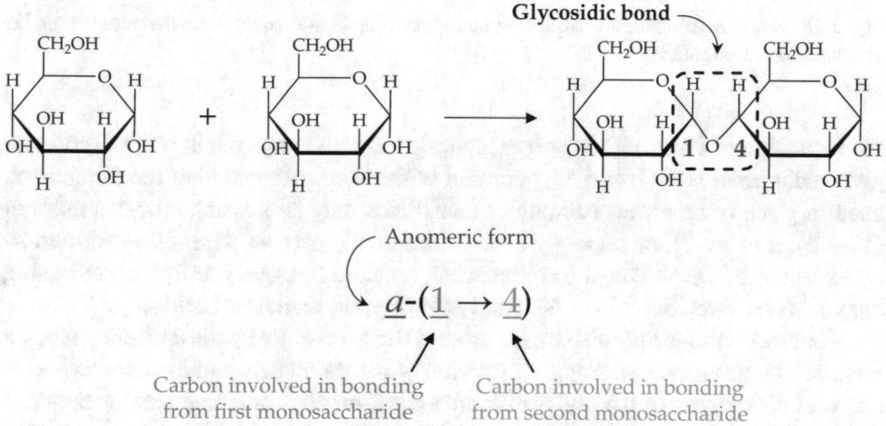

Fig. 2.17 Glycosidic bond formation between two glucose molecules and notation of the bond. In this example, it is an α-(1 → 4) glycosidic bond

may also link through a nitrogen (*N*-glycosides) or sulfur (*S*-glycoside) atom. Many natural compounds in foods are glycosides such as flavonoids, steroids, or coumarins. *Steviol glycosides* are responsible for the sweet taste of stevia that is used as a non-calorific sweetener. Hesperidin, naringin, and rutin are flavonoids with a bitter taste and may need to be removed from the product (e.g., grapefruit juice).

Glycosidic bond notation involves *three elements*: (i) the anomeric form of the sugar that carries the hemiacetal group, (ii) the number of the carbon atom involved in the bond from the first monosaccharide, and (iii) the number of the carbon atom that is involved in the bond from the second monosaccharide. When a disaccharide or polysaccharide has α-(1 → 4) links, the constituent monosaccharides are in the alpha (α) anomeric form. Additionally, the first monosaccharide connects to the second with the C-1 carbon and the second monosaccharide connects to the first with the C-4 carbon (Fig. 2.17). In the case of β-(1 → 4), the only difference is that the monosaccharides are in the β anomeric form. An α-(1 → 6) indicates that the link in the second monosaccharide occurs at C-6, and this linkage is responsible for

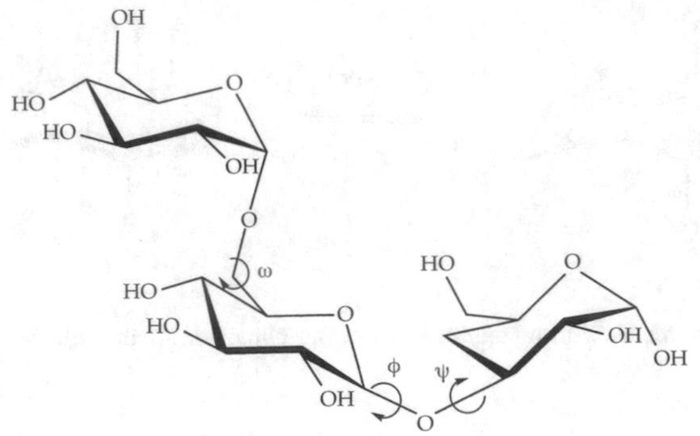

Fig. 2.18 Rotation of monosaccharides around glycosidic bonds. φ, ψ, ω are the possible angles of monosaccharide rotation

the formation of *branching* in polysaccharides. Another equivalent way of writing a glycosidic bond is α-(1,4), i.e., a comma without space separating the numbers is used in place of an arrow. Although other bonds may be possible, for example, α-(1 → 3), β-(1 → 2), or β-(1 → 3), the three bonds α-(1 → 4), α-(1 → 6), and β-(1 → 4) are by far the most prominent. Glycosidic linkages are hydrolysed using enzymes (see Sect. 3.4.3.1) or heat and acid to yield monosaccharides.

Monosaccharides may also *rotate* around the glycosidic bonds until they reach a state of minimum energy. Angle of rotation φ is located between the anomeric carbon and the oxygen of the glycosidic linkage of the first monomer, and ψ between the oxygen of the glycosidic linkage and the non-anomeric carbon of the second monomer (Fig. 2.18). The introduction of branching at C-6 gives one more possible angle of rotation (ω) about the C-5 and C-6 bond. Branching and rotation are critical as the molecule gets bigger (e.g., polysaccharides), as they control their functional properties (e.g., viscosity or gelation).

Carbohydrate Determination

In the *phenol-sulfuric acid method*, sugars react with concentrated sulfuric acid to produce furfural derivatives. Addition of phenol results in products with a yellow colour that can be determined colourimetrically.

$$\text{Sugar} \xrightarrow{\text{Sulfuric Acid}} \text{Furfural derivatives} \xrightarrow{\text{Phenol}} \text{Yellow colour}$$

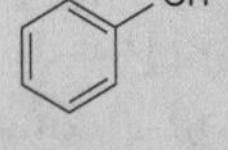

The *m-hydroxydiphenyl method* used to determine *total uronic acids* is also based on the reaction of sugars in concentrated sulphuric acid followed by the formation of pink coloured complexes with *m*-hydroxydiphenyl reagent that can be determined colourimetrically.

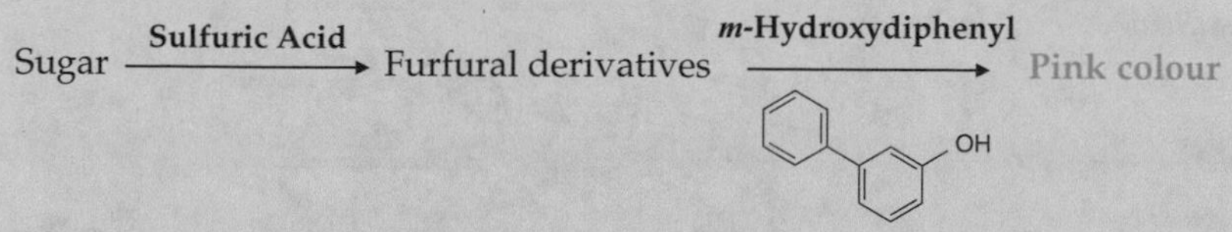

Sugar —Sulfuric Acid→ Furfural derivatives —*m*-Hydroxydiphenyl→ Pink colour

Reducing sugars can be determined with the *Fehling's* or *Benedict's* reagents that contain Cu^{2+}, which is reduced to Cu^+. The reactions can be used in both qualitative and quantitative (with suitable modifications) analysis of reducing sugars. Polysaccharides are determined with quantification of monosaccharides that result after their complete hydrolysis, followed by an appropriate chromatographic method.

2.4   Disaccharides-Oligosaccharides

As has already been mentioned, the result of joining two monosaccharides is the formation of a disaccharide. *Maltose* consists of two glucose molecules while retaining the reducing hemiacetal at the C-1 that is not involved in the glycosidic bond (Fig. 2.19). *Lactose* that consists of glucose and galactose also retains the anomeric C-1. Lactose intolerant individuals cannot break the glycosidic bond of lactose during digestion, resulting in gastrointestinal discomfort. Lactose-free milk production involves hydrolysis of lactose, most frequently using the enzyme *lactase* to produce a mixture of glucose and galactose. As a result, lactose-free milk is slightly sweeter than regular milk because of the presence of glucose. *Sucrose* consists of glucose and fructose linked by an α-(1 → 2) glycosidic bond. Hydrolysis of sucrose to its constituent sugars (i.e., glucose and fructose) using the enzyme *invertase* or heat-and-acid produces a syrup known as *invert sugar*. Invert sugar is sweeter than table sugar and is more difficult to crystallise, making it an ideal sweetener for the confectionery and baking industry. It is called "invert sugar" because the

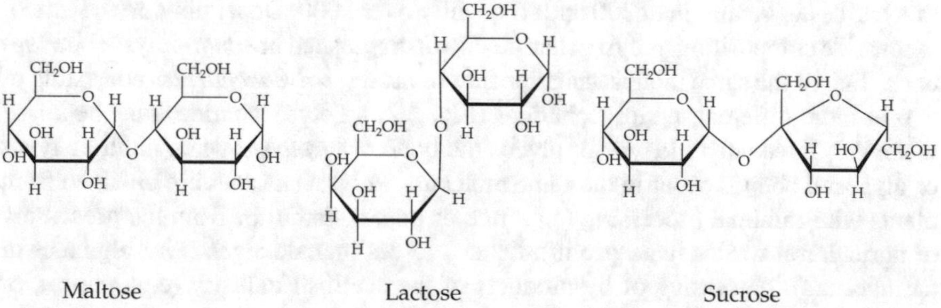

Maltose   Lactose   Sucrose

Fig. 2.19 Common disaccharides found in food. Sucrose is a non-reducing disaccharide

Fig. 2.20 Common oligosaccharides found in foods

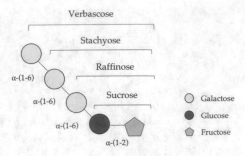

mixture of glucose and fructose that is produced during hydrolysis shifts the *optical rotation* to the opposite direction than that of sucrose from +66.5° (pure sucrose) to −19.7° (for fully hydrolysed sucrose).

Oligosaccharides have between two and twenty monosaccharides linked with glycosidic bonds, and very few of them occur naturally (Fig. 2.20). *Raffinose* (trisaccharide), *stachyose* (tetrasaccharide) and *verbascose* (pentasaccharide) are some commonly occurring oligosaccharides of beans that are non-digestible and responsible for flatulence. *Fructo-oligosaccharides* are naturally occurring and consist of multiple units of fructose, frequently termed *fructans*. Foods rich in fructans are asparagus, banana, garlic, leek, onion, and Jerusalem artichoke. Commonly occurring fructans are *inulin* and *levan* with fructose units linked by β-(2 → 1) or β-(2 → 6) bonds. They can be used as *prebiotics* inducing the growth of beneficial bacteria in the gut, as fat replacers, or texture modifiers.

2.5 Polysaccharides

Polysaccharides are *monosaccharide polymers*, as several monosaccharide units are linked *via* glycosidic bonds. The large molecule, frequently termed *polysaccharide chain*, resembles a string of beads. The number of monosaccharides comprising the chain is termed *degree of polymerisation* (DP). For example, when the DP is 200, then the polysaccharide consists of 200 monosaccharides. Usually, the DP of polysaccharides is greater than 200 and frequently over 3000. Depending on the monosaccharide composition, polysaccharides are distinguished in *homo-polysaccharides* consisting of the same monosaccharide unit or *hetero-polysaccharides* consisting of two or more different monosaccharides (Fig. 2.21). Polysaccharides may be *linear* when they have only one type of glycosidic bond or *branched* when multiple types of glycosidic bonds occur in the same molecule. Polysaccharides are obtained from plants with minimal processing (e.g., rice or potato starch) or from the processing of agricultural wastes (e.g., pectin). Other sources include algae (e.g., alginates or carrageenan), processing of by-products of the shellfish industry (e.g., chitin), or microbial fermentation (e.g., xanthan or gellan).

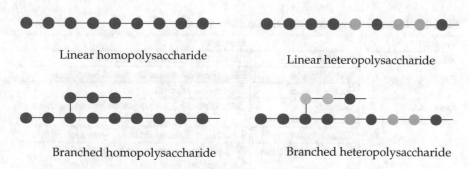

Fig. 2.21 Types of different polysaccharide structures. See text for explanation

Although in nature there are numerous monosaccharides, the number of those found in polysaccharides is relatively small. Common monosaccharides include glucose and mannose that form the backbone of some of the most important commercial polysaccharides. Other sugars or sugar acids such as galactose, xylose, arabinose or galacturonic, guluronic and mannuronic acids are found in industrially relevant polysaccharides. However, the type of linkages, isomeric forms, functionalisation of monosaccharides, branching, and periodicity of the monomers in the backbone results in significant structural variability and diversity in functionality. These modifications are, for example, methylation or acetylation at various positions, the presence of sulfate or other functional groups or differences in the anomeric type of monosaccharides that make up the polysaccharide. A notable example is that of amylose and cellulose that both consist of glucose. Glycosidic linkages between glucose units in amylose are α-(1 $\rightarrow$ 4), whereas in cellulose are β-(1 $\rightarrow$ 4), resulting in different functional properties not only within the plant (structural *vs* source of energy) but also when they are used as food ingredients.

Polysaccharides are essential food ingredients, and it is difficult to find a food that does not have a polysaccharide present either naturally or as an additive. The term "gum" is also used to describe polysaccharides and is usually reserved for non-starch polysaccharides, for example, guar gum, gum Arabic, or xanthan gum. Many terms are available in the literature to describe polysaccharide functionality, such as bulking agents, emulsion stabilisers, fat replacers, gelation agents, thickening agents, viscosity enhancers, water-binding agents, or texture modifiers. Regardless of application specifics, polysaccharide functionality falls into two main classes: *structure formation* and *structure stabilisation* (Table 2.1). When used as structure-forming agents, they form *gels* that hold a large amount of water and create a soft-solid three-dimensional structure. This functional property is used to hold the parts of the formulation together (e.g., jams, onion rings or batter in chicken nuggets) and provide essential textural characteristics to the product. When they are used as *stabilisers*, they *increase viscosity*, thus preventing physical separation (e.g., creaming or precipitation). This ensures the processing of a uniform formulation that otherwise would be problematic (e.g., ice cream mix before freezing) or prolonging the shelf life of products (e.g., suspension of particles in salad dressings). Frequently, polysaccharides have more than one role in the same formulation. For instance, in

Table 2.1 Examples of polysaccharide functionality in foods

Property		Examples
Structure formation	Gelation	Gelation in onion-ring formation
	Emulsification	Emulsification of flavour oils in beverages
	Encapsulation	Conversion of oils into powder
Structure stabilisation	Viscosity enhancement	Prevention of phase separation in salad dressings
	Water retention	Prevention of syneresis in dairy desserts
	Texture modification	Changes thickness perception in low-fat formulations

salad dressings, they may prevent phase separation of oil droplets and provide texture modification with an increase in viscosity. In most cases, they interact with other food ingredients (e.g., proteins, lipids, or cations), resulting in a complex interplay of interactions that makes food formulation with polysaccharides quite a challenging task. As a result, understanding of polysaccharide chemical structure is the very first step before application.

2.5.1 Starch

Starch is the carbohydrate source in plants found in tubers, fruits, and seeds. It is widely used in food and non-food industries (e.g., paper, chemical, or cosmetic), and there is an extensive selection of starches from various sources such as corn, wheat, rice, potato, or cassava (tapioca). Naturally found starch that has not been further processed is termed *native*. However, native starches should be physically, chemically, or enzymatically changed to obtain technological consistency and are called *modified starches*.

Starch is found in the *starch granule* and is a mixture of two polysaccharides: *amylose* and *amylopectin*. Amylose and amylopectin are glucose homopolysaccharides and differ only in branching. Glucose units in amylose connect with α-(1 $\rightarrow$ 4) glycosidic bonds (Fig. 2.22a). The spatial positioning of glucose units gives amylose a *helical shape*. The core of this structure contains hydrogen atoms and is hydrophobic, whereas the -OH groups are positioned on the outer part of the helix making it hydrophilic. Amylose may form complexes with hydrophobic molecules such as lipids or flavours that are entrapped in the hydrophobic core of the molecule. Iodine may also bind into the core and forms the basis of amylose determination in starches.

Amylopectin is a large branched molecule having both α-(1 $\rightarrow$ 4) and α-(1 $\rightarrow$ 6) linkages (Fig. 2.22b). The presence of α-(1 $\rightarrow$ 6) creates branches forming three types of chains. The chain that carries the only reducing end is called *C-chain* (Fig. 2.23a). Branches stemming from the C-chain are termed *B-chains*, whereas those originating from B-chains are called *A-chains*. A-chains do not have any

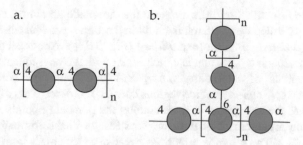

Fig. 2.22 Glycosidic bonds in (**a**) amylose, and (**b**) amylopectin. *n* is the number of *repeating units* shown within the square brackets. In amylose, for example, if *n* = 100, the polysaccharide consists of 200 glucose molecules because within the brackets are two glucose molecules. Blue circle: glucose

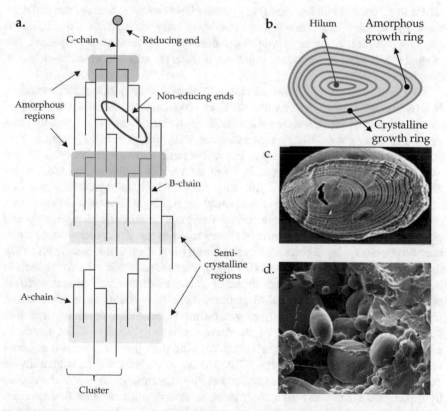

Fig. 2.23 (**a**) Structure of amylopectin (see text for description), (**b**) schematic representation of a starch granule exhibiting growth rings consisting of alternating regions of semi-crystalline and amorphous regions, (**c**) microscopy image of growth rings in a starch granule, and (**d**) microscopy image of starch granules embedded in bread dough. The fibrous structures are gluten strands and starch granules have different sizes. Microscopy image (**c**) is reproduced with permissions from Pilling and Smith, (2003), *Plant Physiology*, 132, 365–371. Microscopy image (**d**) is reproduced with permissions from Kontogiorgos, Goff, Kasapis, (2008), *Food Hydrocolloids*, 22, 1135–1147

branches. Starches with very little amylose that are made up almost exclusively of amylopectin are called *waxy*. Starch is found in the *starch granule* where amylopectin forms an *ordered structure* (*crystalline*) (Fig. 2.23a). Amylopectin molecules orient radially around a point that is called *hilum* with the consequent formation of concentric regions of alternating *amorphous* and *semi-crystalline* structures (Fig. 2.23b, and c). Semi-crystalline regions consist of ordered parts of amylopectin chains, whereas the amorphous regions contain the branching points and amylose. The alternating arrangement of these regions results in the formation of *growth rings* (Fig. 2.23c). The hilum is important because water enters from this opening and results in starch granule *swelling* facilitating *gelatinisation*. Starch granules have different shapes and sizes that determine their swelling capacity and other physical properties. Large (>10 μm) and lenticular granules (like a lens) called *A-type starch granules* are found in cereal endosperms. Small (<10 μm) and spherical, called *B-type starch granules*, have a lower swelling capacity and are found in tubers (e.g., potato) and legumes (e.g., peas). Native starch granules have different sizes, i.e., starch is not composed of granules of only one size. As a result, starch granules exhibit a *size distribution* that influences their functional properties. For example, it is easy to see starch granules of different sizes in wheat flour bread-dough (Fig. 2.23d).

As mentioned earlier, industrial starch is classified into native and modified. Native starches exist in their natural form, as obtained from their source. Some of the most common native starches are corn (maize), rice, wheat, potato, tapioca (cassava), and waxy maize. *Waxy maize* and other waxy native starches generally contain less than 2% amylose, whereas the rest contain somewhere between 15 and 30%. *High-amylose* starches contain more than ~30% amylose and are usually unsuitable for food applications due to their very high retrogradation rate (see below). Native starches differ in crystal structure, pasting characteristics, and minor components present within amylose and amylopectin such as phosphate esters, phospholipids, or proteins. Because of these variations and their inconsistent nature, they do not always have good pasting or gelling properties, and it is difficult to use them in food formulations. With modification, starch functionality improves, including resistance to heat treatment, shear, or acidic environments. Modification of starches using chemical or physical methods includes *crosslinking*, *derivatisation*, *acid thinning* (*depolymerisation*), *pre-gelatinisation*, *oxidation,* or a combination of them. Starch modified with any of the above methods is called *modified starch*. In *crosslinked starch*, linking reagents react with the -OH groups from two adjacent molecules and merge two chains. This process adds acid and heat stability to starches, improves viscosity and water holding capacity. In *derivatised starches* hydroxyl groups are modified usually using an esterification or etherification reaction by adding bulky groups to the structure (see Section 2.3). Derivatisation reduces the tendency of chains to associate, reduces retrogradation, increases water binding capacity and the tendency of the paste to gel. The introduction of hydrophobic groups imparts emulsification properties to starch (e.g., esterification with Octenyl Succinic Anhydride (OSA-starch)). Acid hydrolysis yields *acid thinned* starches with smaller chains and a concomitant reduction in viscosity. The reaction of starch with oxidising agents, usually hypochlorite, results in *oxidised starches*. Oxidised starch exhibits low viscosity due to depolymerisation and improved stability of starch dispersions.

Native starch granules are not soluble in cold water but may absorb water through the hilum and swell. When heated in water, granules go through an event called *gelatinisation*. Gelatinisation is the disruption of the molecular order within the granules resulting in loss of crystallinity, amylose leaching, and starch solubilisation. Several factors influence the initial temperature and the range over which it occurs, such as starch type, starch-to-water ratio, pH, or concentration. Gelatinisation occurs over a *temperature range* usually within ~55–95 °C and the result is a *starch paste*. The above events are followed using the *pasting curve* that shows the viscosity of starch dispersion as a function of temperature (heating and cooling) (Fig. 2.24). The viscosity of the dispersion is initially low, but during the swelling stage, it increases rapidly as the starch granules swell. After peak viscosity has been reached, the granule bursts and the viscosity drops. Following completion of gelatinisation, cooling of starch paste results in *retrogradation* that refers to the *recrystallisation of amylose and amylopectin*. The term gelatinisation is commonly confused with the term *gelation* but they refer to completely different physicochemical events. Gelatinisation occurs only in starch, as described above, but gelation is a general property of polysaccharides, proteins, lipids, or their mixtures to form a

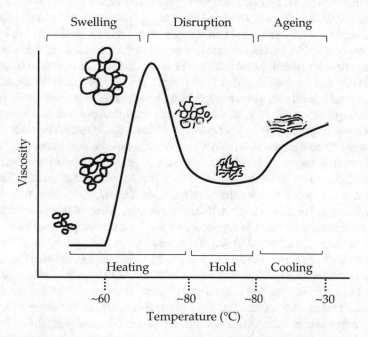

Fig. 2.24 Gelatinisation events and the pasting curve of starch. Starch granules swell and eventually burst, and viscosity decreases. In cooling, amylose and amylopectin aggregate in a process known as retrogradation

three-dimensional network (a *gel*) that can hold large amounts of water (i.e., a *hydrogel*) or oil (i.e., an *oleogel*).

Amylose recrystallisation proceeds much faster than that of amylopectin and is the major reason for the observed quality defects in starch-containing products. For example, retrogradation is responsible for bread and bakery products staling and loss of viscosity and precipitation in soups and sauces containing starch. Retrogradation is a common defect of *gluten-free* formulations that are heavily based on starch and results in rapid texture deterioration and short shelf life of the product. Waxy starches, i.e., starches that contain very low amounts of amylose, do not show appreciable retrogradation. The utilisation of modified starches may eliminate or slow down retrogradation to such an extent that it has a negligible impact on the quality and shelf life of the starch-containing product. Native starches do not form pastes in cold water and therefore require heating and go through the gelatinisation process, as described above. *Pregelatinisation* is a type of *physical modification*, i.e., without the use of chemical reagents that results in starch that can be dispersed in cold water. *Pregelatinised starch* is formed after heating starch in water until complete gelatinisation, followed by drying and grinding to a fine powder. It usually forms paste in cold water at concentrations between 1 and 5%. This type of starch is used in instant powders, pie fillings, soup mixes, salad dressings, confectionery, and meat products. Nutritional value is the same as that of the original starch.

Resistant starch (*RS*) is starch that cannot be digested in the small intestine of healthy individuals. It is divided into four types, namely, RS 1 to 4, depending on its properties. RS1 is *physically inaccessible* that digestive enzymes cannot come into contact, such as in the intact wholegrain unprocessed cereals, seeds, nuts, or legumes (e.g., oats, rye, wheat, barley, semolina, corn, linseed, or sesame). RS2 is starch that *digestive enzymes cannot break* and occurs in its natural form, such as in bananas, unripe fruits, and legumes (e.g., lentils or beans). RS3 forms when starch-containing foods are *cooked and cooled again*, as heating and cooling may form resistant starch due to retrogradation. Finally, RS4 starches are *modified starches that resist digestion* because the glycosidic bonds cannot be cleaved by digestive enzymes.

A group of consumers is sensitive to a certain type of prolamins (i.e., *coeliac disease*) mainly found in the gluten of wheat flour. Consumption of wheat-based products from these individuals results in gastrointestinal discomfort. The only treatment is to eliminate gluten from their diet. As a result, gluten-free bakery products require a different approach in their formulation. Most of them are formulated with a combination of starches supported by the inclusion of polysaccharides (e.g., xanthan or carboxymethyl cellulose) and plant proteins (e.g., pea or soy). The major problems of gluten-free products are low product volume, dense structure, and texture losses during storage. The first two result from the absence of gluten responsible for the desirable structure formation of bakery products. The third results from the rapid retrogradation of starches used in the formulations. Consequently, potato, tapioca, waxy maize, or modified starches are preferred, as they generally have low retrogradation rates. The inclusion of minor ingredients (e.g., lipids, salt, monosaccharides, enzymes etc.) is also used to retard retrogradation even further and improve texture and shelf life. Because of these formulation requirements, gluten-free

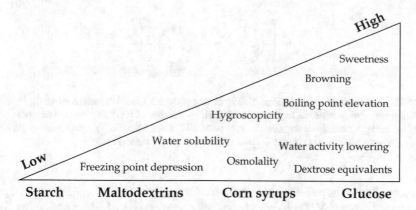

Fig. 2.25 Changes in the functional properties of glucose-based ingredients with the increase in dextrose equivalents (from starch to glucose). The intensity of each property increases from left (low) to right (high)

products are usually higher in carbohydrates, fat, and NaCl than their regular counterparts.

Maltodextrins are oligosaccharides with up to 20 glucose units linked with α-$(1 \rightarrow 4)$ glycosidic bonds, whereas *glucose syrups* are starch hydrolysates with quite different properties. Both maltodextrins and glucose syrups are products of starch hydrolysis. The extent of hydrolysis is measured in *dextrose equivalents* (DE). DE is a measure of the reducing sugars in the hydrolysate, and it ranges from 0 to 100%. Pure glucose has a DE of 100 (i.e., 100% reducing sugars), whereas starch has 0% (i.e., 0% reducing sugars). Maltodextrins with a DE of 10 would have 10% of the reducing power of glucose. We need to know the DE value, as it affects the functional properties of the resulting maltodextrins and glucose syrups (Fig. 2.25). Maltodextrins may be used as bulking or viscosity enhancing agents or encapsulating various components (e.g., colours, flavours, or oils). Glucose syrups are used as sweeteners, texture modifiers, browning agents, or sugar crystallisation inhibitors in confectionery products.

2.5.2 Carrageenan

Carrageenan is a linear, sulphated galactan composed of alternating 3-linked β-D-galactose and 4-linked α-D-galactose or 4-linked 3,6-anhydro-α-D-galactose, thus forming its disaccharide repeating unit and is obtained from seaweeds (Fig. 2.26). The most industrially relevant types of carrageenan are κ-, ι-, and λ-carrageenan that have one, two, or three ester-sulphate groups (O-SO$_3^-$) per repeating disaccharide unit, respectively. Carrageenan may form gels after heating to about 80 °C followed by cooling to approximately 40–60 °C. At high temperatures, they form chains resembling "boiled spaghetti" termed *random coils*. On cooling, they undergo

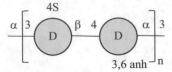

Fig. 2.26 Structure of κ-carrageenan. Yellow circle: α-D-galactose, 4S: sulfate group on C-4, 3,6 *anh*: 3,6-anhydro-α-D-galactose, D: D-enantiomer, *n*: number of repeating units. Structures of ι- and λ- carrageenan are very similar but with different amounts of ester-sulfate content. κ-Carrageenan has one, ι- two, and λ- three sulfate groups per disaccharide

a process known as *coil-to-double helix conformational transition* resulting in a gel. Potassium cations (K⁺) help with helix-helix aggregation of adjacent chains. Gels are thermoreversible and melt on heating. Because of the presence of ester-sulfate that gives a negative charge to the polysaccharide at the pH of milk, carrageenan can interact strongly with positively charged amino acids of κ-casein through ionic interactions. These interactions result in structure formation that finds applications in many dairy products such as chocolate milk (e.g., cocoa particle stabilisation), dairy desserts, or ice creams.

2.5.3 Alginates

Alginates is the term used for the salts of alginic acid, but it also refers to all the derivatives of alginic acid and alginic acid itself. Alginates are structural compo-nents of brown seaweed present in the form of divalent salts of alginic acid and form the intercellular gel matrix. Alginates are linear polysaccharides that are composed of β-D-mannuronic acid (M) and α-L-guluronic acid (G) linked *via* (1 → 4) glyco-sidic linkages (Fig. 2.27). Accordingly, the alginate backbone consists of sequences of blocks of mannuronic acid (*M-blocks*) or guluronic acid (*G-blocks*) and regions of alternating sequences (e.g., MG, MMG, GGM). The M-blocks are linked *via* β-(1 → 4) linkages adopting $^4C_1$ chair conformation that imparts flexibility to the chain. In contrast, G-blocks are rigid structures due to the $^1C_4$ conformation of gulu-ronate residues linked *via* α-(1 → 4). As a result, the stiffness of the blocks on the backbone follows the order (most rigid) GG > MM > MG (least rigid). The most prominent physical property of alginates is the selective binding of divalent and multivalent cations determining their ability to form gels. Gelation of alginates is described by the *egg-box model*, according to which alginate chain–chain interac-tions are induced by the presence of cross-linking divalent cations, most frequently Ca²⁺. Alginates may be used in dairy (e.g., creams and whipped creams, or pro-cessed cheese) or in restructured food products (meat, fruits, vegetables, or fish) as well as in the pharmaceutical industry (e.g., antacids or wound dressings).

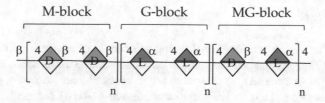

Fig. 2.27 Structure of alginates. Green/white: D-mannuronic acid, orange/white: L-guluronic acid, *n*: number of repeating units

Alginate Gelation
We have discussed that alginates may form gels with a mechanism described as the *egg-box model*. But how did we come up with this term? Guluronic acid units of two alginate chains that run antiparallel to each other create "holes" where Ca^{2+} embeds. This formation creates ionic *cross-links* between alginate chains, as the positively charged calcium cations interact with the negatively charged carboxyl groups. A very similar mechanism also occurs in the gelation of LM-pectins. This chain formation resembles an egg box with the calcium being the egg and the chains the box!

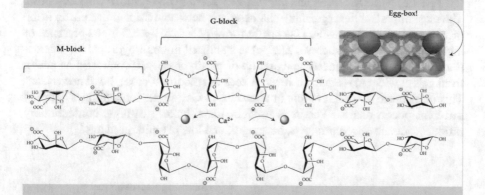

Polysaccharides have different ways of creating gels, but they are all based on the way chains interact. The forces that are involved are ionic (e.g., alginate gels), hydrogen bonds (*β*-glucan gels), or hydrophobic interactions (e.g., methylcellulose gels). Quite frequently, a combination of the above forces coincides. The quality of the final gel, such as its rigidity or water holding capacity, is highly specific to the type of polysaccharide and depends on many factors, including the concentration of polysaccharide, pH, temperature, or molecular weight, just to name a few.

2.5.4 Pectin

Pectin is a heteropolysaccharide belonging to a family of covalently linked galacturonic acid-rich polysaccharides and is obtained commercially from citrus fruits (e.g., oranges or lemons), apples, and sugar beets. Pectin primarily consists of *homogalacturonan* (HG) and *rhamnogalacturonan-I* (RG-I) segments. The intra- and intermolecular interactions between these two segments control its functional properties (Fig. 2.28). HG is the most abundant and is composed of long chains of linear 1 → 4 linked α-D-galacturonic acid residues (~200 units). Some of the carboxyl groups are methyl-esterified at the C-6 position and acetyl-esterified at the O-2 and O-3 positions of the galacturonic acid, depending on plant species. RG-I is composed of the repeating disaccharide galacturonic acid and rhamnose [α-(1 → 2)-D-galacturonic acid–α-(1 → 4)-L-rhamnose]$_n$ where n can be greater than 100. RG-I segments usually have branches of galactose or arabinose that may also carry a significant amount of protein attached to them. Because of branching, the RG-I segment of the molecule is also known as the *hairy region,* whereas the HG segment that lacks branching is called the *smooth region. Degree of esterification* (DE) is the percentage of esterified galacturonic acid units out of the total number of galacturonic acid units present in pectin. Pectins are divided into two commercially important groups with distinct functional properties: *high methoxy pectins* (HM-pectin) with DE > 50% and *low methoxy pectins* (LM-pectin) with DE < 50%. The most important distinguishing characteristic between HM and LM pectin is the mechanism of gel formation. HM pectin forms gels at pH < 3.5 in the presence of high concentrations of sucrose. The gel is stabilised through hydrophobic interactions due to sucrose-induced dehydration of pectin chains. In contrast, LM pectin form gels at wider pH range, however, it does not require sucrose, but it needs Ca^{2+}. This mechanism is similar to that of alginates, as Ca^{2+} forms ionic bridges between different pectin chains. Pectin is used in jams, bakery fillings, confectionery products, tomato-based products, beverages, and low pH milk products.

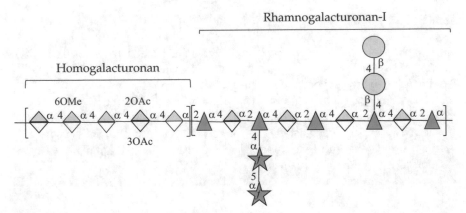

Fig. 2.28 Structure of pectin. Yellow/white diamond: galacturonic acid, green triangle: rhamnose, green star: arabinose, yellow circle: galactose, OMe: methyl ester, OAc: acetyl ester, *f:* furanose

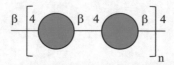

Fig. 2.29 Structure of cellulose. The only difference between amylose and cellulose is that cellulose has glucose in the β anomeric configuration. This results in substantial differences in the chemical and physical properties between the two. Blue circle: glucose

2.5.5 Cellulose

Cellulose is the major cell-wall component of plants and is composed of D-glucose connected *via* β-(1 → 4) glycosidic bonds (Fig. 2.29). Cellulose is insoluble in water and indigestible by human digestive enzymes making it the major component of *dietary fibre*. Although cellulose is insoluble in water with appropriate substitutions can be converted into a water-soluble polysaccharide. These water-soluble polysaccharides are collectively known as *cellulose derivatives*. Examples are *carboxymethyl-cellulose* (CMC), where a carboxymethyl group is attached after etherification and can be used to control ice crystal growth in ice creams. *Methylcellulose* (MC) is cold-water-soluble and *forms thermoreversible gels on heating* making it insoluble in hot water. MC gels are stabilised by hydrophobic interactions between methyl groups belonging to different chains. Many vegetarian and vegan "burgers" cannot be formed without MC or similar cellulose derivatives. A plant-protein-based "burger" disintegrates on heating leaving behind an undesirable structure with poor sensory attributes because plant proteins do not have the same structuring ability as those from meat (e.g., myosin and actin). In the presence of MC, a thermoreversible gel is formed on heating (i.e., cooking), holding the formulation together. On cooling to eating temperature (~30 °C), the gel melts and returns to a viscous liquid state, and the solid gel is not detectable during consumption. Using the same principle, MC is used in a range of products that are deep-fried before consumption (e.g., chicken nuggets, fish fingers, etc.) to preserve moisture and desirable textural properties. *Microcrystalline cellulose* (MCC) is partially depolymerised cellulose that is water-soluble and can be used as a non-caloric filler in various food products.

2.5.6 Galactomannans

Galactomannans are seed polysaccharides commercially obtained mostly from *guar* and *carob* (locust bean) having a backbone of β-(1 → 4)-linked mannose with galactose branches linked with α-(1 → 6) glycosidic bonds (Fig. 2.30). The

Fig. 2.30 Structure of
galactomannans. Green:
mannose, yellow: galactose

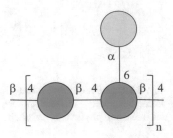

distinguishing feature between different galactomannans is the mannose-to-galac-
tose ratio (M/G), which is characteristic for each source. The ratio ranges approximately between 2:1 or 4:1 for guar or carob gum, respectively. Galactomannans
form viscous solutions at low concentrations and can be used together with xanthan
or carrageenan to improve the stabilising or gelling properties of food formulations.
They are commonly used in ice cream to prevent ice recrystallisation on storage, in
salad dressings to improve viscosity and water holding capacity, or in soft cheese
products to improve spreading.

2.5.7   Gum Arabic

Gum Arabic (*Acacia gum*) is a tree exudate obtained from Acacia tree species
(*A. senegal* or *A. seyal*). It is a highly branched heteropolysaccharide with β-(1 → 3)
galactose forming the backbone and branches at the C-6 position consisting of
galactose, arabinose, rhamnose and glucuronic acid. This structure is attached to a
protein component that plays a key role in its functionality (Fig. 2.31). The presence
of protein makes gum Arabic the most technologically important polysaccharide-
based emulsifier. It is highly water-soluble and forms low viscosity solutions even
at very high concentrations (20–30%). Because of its low viscosity, it can emulsify
flavour oils (e.g., orange oil) in the confectionery and beverage industry (e.g., fruit
flavoured beverages) without giving a slimy texture during consumption of the
beverage.

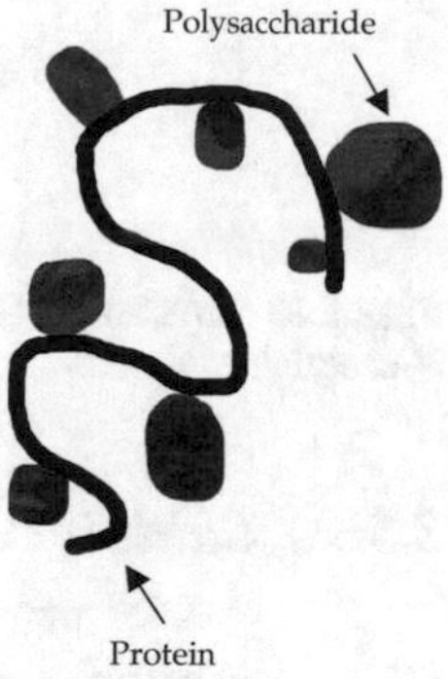

Fig. 2.31 Schematic representation of gum Arabic structure. Gum Arabic has a very compact shape allowing it to have a low viscosity at high concentrations

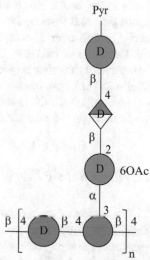

Fig. 2.32 Structure of xanthan. Blue circle: glucose, green circle: mannose, blue/white diamond glucuronic acid, Pyr: pyruvic acid, OAc: acetyl group

2.5.8 Xanthan

Xanthan is a heteropolysaccharide produced by the bacterial fermentation of *Xanthomonas campestris*. It has a cellulose backbone, and every alternate glucose residue has a three-sugar side chain consisting of two mannose residues with a glucuronic acid residue between them. The mannose that resides nearest to the main chain may carry a C-6 acetyl group, and the terminal mannose may carry a pyruvate group between C-4 and C-6 (Fig. 2.32). If this structure description seems to be complex, you can consider xanthan as cellulose with trisaccharide branches. This structure results in *very stiff chains* with very limited capacity to fold when they are in solution. As a result of its stiff structure, xanthan forms highly viscous solutions at *very low concentrations*, as low as 0.01% depending on the formulation. It is used as a stabiliser in emulsions (e.g., mayonnaise, salad dressings etc.) or bakery products in gluten-free formulations.

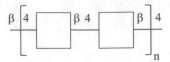

Fig. 2.33 Structure of chitin. Chitin has very limited food applications. White square: *N*-acetylglucosamine

2.5.9 Chitin

Chitin consisting of *N*-acetylglucosamine linked *via* β-$(1 \rightarrow 4)$ bonds, is an amide and acetylated derivative of glucose (Fig. 2.33). It is the primary component of cell walls in the exoskeletons of arthropods, such as crustaceans (e.g., shrimp shells, squid pen) and insects. Deacetylation of chitin in alkaline environments removes acetyl groups from *N*-acetylglucosamine to form D-glucosamine, which results in the formation of *chitosan*. Chitosan is the *only positively charged polysaccharide* compared to the other polysaccharides that are either neutral or negatively charged. Chitin and chitosan belong to the group of polysaccharides known as GAGS (*glycosaminoglycans*). Chitosan has very limited commercial uses in foods, but it finds applications in the pharmaceutical and biomedical industries.

2.5.10 Dietary Fibre

The definition of the term *dietary fibre* (DF) keeps changing as new scientific evidence is presented in chemistry, nutrition, and human physiology. Depending on the emphasis, they may be classified based on their ability to ferment in the large intestine or their solubility in aqueous buffers. The common characteristic of DF is that they are not hydrolysed by the endogenous enzymes in the small intestine of humans and exhibit beneficial health effects. Regardless of the definition, *dietary fibres are cellulose, hemicellulose, pectin, resistant starch, and lignin*. The structure and major properties of cellulose, pectin and resistant starch have been described in the previous sections. *Hemicelluloses* are heteropolysaccharides with quite diverse structures found in the plant cell walls. The term "hemicelluloses" may create confusion, and one may attempt to find structural similarities with cellulose. However, hemicelluloses have a different structure than cellulose and depending on the monosaccharide that forms the backbone of the molecule they may be *xylans* (e.g., arabinoxylans), *mannans* (e.g., glucomannans), or *glucans* (e.g., β-glucan). Hemicelluloses can be linear (e.g., β-glucan) but frequently have branching points. For example, in arabinoxylans, the backbone consists of xylose and the branches of arabinose. *Lignin* is also part of the cell wall, but it is not a polysaccharide and is not fermented in the colon. Lignin is a highly heterogeneous polymer of oxygenated phenylpropane derivatives, including *coniferyl*, *sinapyl* and *p-coumaryl* alcohols

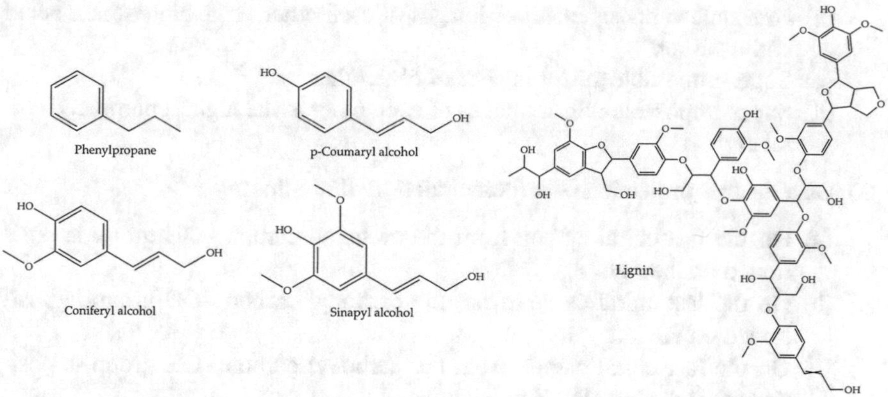

Fig. 2.34 Lignin is a polymer of phenylpropane derivatives (e.g., coniferyl, sinapyl, and *p*-coumaryl alcohols)

(Fig. 2.34). Lignin has very diverse structures and molecular weight and does not have a single chemical formula.

Dietary fibres may be classified into two general groups based on their water solubility: *soluble dietary fibre* (SDF) and *insoluble dietary fibre* (IDF) with quite different properties and challenges during food formulation. For example, SDFs increase viscosity of the formulation in contrast to IDFs that have limited capacity to enhance viscosity. Generally, formulating foods with ingredients enriched in IDF creates technological problems stemming from their limited hydration, inhomogeneous dispersion, or poor interactions with other food compounds. Frequently, final products have unacceptable textures and limited shelf life compared to their low IDF counterparts. Consequently, understanding the chemical structure of polysaccharides and how they interact with other food compounds is important in high DF food formulation.

2.6 Learning Activities

2.6.1 Multiple-Choice Questions – Monosaccharides

1) An aldohexose is:

   (a) A monosaccharide with aldehyde group and five carbons
   (b) A monosaccharide with ketone group and six carbons
   (c) A disaccharide with aldehyde group and five carbons
   (d) A monosaccharide with aldehyde group and six carbons

2) Two structures that have a chiral carbon in their structure are:

   (a) Non super-imposable mirror images of each other

(b) Non super-imposable mirror images of each other at the highest numbered chiral carbon

(c) Super-imposable mirror images of each other

(d) Super-imposable mirror images of each other at the highest numbered chiral carbon

3) On a Fischer projection D- carbohydrates are those that:

(a) On the last chiral carbon from the carbonyl carbon –OH group is positioned on the right

(b) On the last chiral carbon from the carbonyl carbon –OH group is positioned on the left

(c) On the last chiral carbon from the carbonyl carbon –OH group is positioned on the left ONLY in hexoses

(d) On the last chiral carbon from the carbonyl carbon –OH group is positioned on the right BUT ONLY in those that have aldehyde group (i.e., glucose)

4) In alpha- and beta- anomers in a Howarth projection, the –OH group is:

(a) below and above the plane of the ring, respectively in D-sugars

(b) above and below the plane of the ring, respectively in D-sugars

(c) below and above the plane of the ring, respectively in amino-sugars

(d) below and above the plane of the ring, respectively in L-sugars

5) The most stable conformation of a pyranose-ring forming sugar is:

(a) The boat conformation

(b) The chair conformation

(c) The cable conformation

(d) The coil conformation

6) A uronic acid is

(a) An acid that results from oxidation of a monosaccharide with urea

(b) An acid that results from the reduction of a monosaccharide with urea

(c) An acid that results from oxidation of terminal -OH of a monosaccharide

(d) An acid that results from oxidation of -OH at C1 of a monosaccharide

7) Aldonic acid is:

(a) a carboxylic acid after oxidation of the ketone group of aldoses

(b) a carboxylic acid after oxidation of the aldehyde group of aldoses

(c) a carboxylic acid after oxidation of the aldehyde group of ketoses

(d) a carboxylic acid after oxidation of the ketone group of ketoses

8) In reducing sugars

(a) The anomeric carbon is free to react, e.g., as in glucose

(b) The anomeric carbon is free to react, e.g., as in sucrose

(c) The chiral carbon is free to react, e.g., as in glucose

(d) The anomeric carbon esterifies, e.g., as in glucose

9) Reduction converts monosaccharides to

(a) Alditols
(b) Uronic acids
(c) Esters
(d) Ethers

10) Glycosidic bond is the bond between two or more:

(a) glycosidic acids
(b) monosaccharides
(c) reducing sugars ONLY in the presence of oxidative enzymes
(d) monosaccharides ONLY when they exist in Howarth projection

2.6.2   Multiple-Choice Questions – Polysaccharides

1) Homo- and hetero- polysaccharides are composed of:

(a) One or several different monosaccharides, respectively
(b) Several different or one monosaccharide, respectively
(c) Linear or branched chains, respectively
(d) Branched or linear chains, respectively

2) Starch consists of:

(a) amylose and pectin
(b) polyamylose and glucose
(c) amylose and amylopectin
(d) amylase and amylopectinase

3) Starch is made up of

(a) Glucose connected with (1,4) and (1,6) glycosidic bonds
(b) Glucose and fructose connected with (1,4) and (1,6) glycosidic bonds
(c) Glucose connected with (1,4) glycosidic bonds
(d) Glucose connected with (1,6) glycosidic bonds

4) Modified starches can:

(a) Slow retrogradation
(b) Act as emulsifiers
(c) Improve pasting characteristics
(d) All of the above

5) Starch gelatinisation is:

(a) The disruption of the molecular order within starch granules
(b) Formation of gel at high starch concentrations
(c) Gelation of modified starches during heating
(d) Formation of viscous solution that improves texture

6) Starch retrogradation is mostly due to

(a) recrystallization of amylose during cooling
(b) recrystallization of amylopectin during cooling
(c) recrystallization of amylose during heating
(d) recrystallization of amylopectin during heating

7) κ-carrageenan:

(a) has ester sulfate content ~50% and does not form gels
(b) has ester sulfate content ~25% and forms gels
(c) has ester sulfate content ~10% and does not form gels
(d) has ester sulfate content ~30% and forms gels

8) The gel formation mechanism of alginates is known as:

(a) Cardboard box model
(b) Corrugated cardboard model
(c) Egg box model
(d) Pizza box model

9) Pectin

(a) consists of guluronic acid and methylated guluronic acid
(b) consists of galacturonic acid and amylopectin
(c) consists of galacturonic acid and methylated galacturonic acid
(d) consists of amylose, amylopectin and galacturonic acid

10) Common cellulose derivative substituent groups are:

(a) Carboxy-ethyl, ethyl, hydroxy-propyl, hydroxy-propyl-ethyl
(b) Carboxy, methyl, hydroxy-, hydroxy-acetyl
(c) Carboxy-methyl, methyl, hydroxy-propyl, hydroxy-propyl-methyl
(d) Carboxy-acetyl, acetyl, hydroxy-acetyl, hydroxy-propyl-acetyl

2.6.3   Short Answer Questions – Further Reading

1) Define all the different stereoisomers that occur in monosaccharides. Draw a pair of structures in each case and identify the differences between them.

2) Draw the Fischer projection of the pairs D-glucose/D-mannose and D-ribose/D-arabinose. Number the carbons correctly from 1 to 5 or 6, identify the chiral carbons and the carbonyl groups.

3) Write the reaction of hemiacetal formation. What is the consequence of this reaction in monosaccharides?

4) Draw the Howarth projection of α-D-glucose and β-D-glucose. Number the carbon atoms correctly from 1 to 6 and identify the anomeric carbon. Try the same with α-L-glucose and β-L-glucose.

5) **Online activity** Find online the following terms and link them to the properties of sugars: "optical rotation", "dextrorotation", and "laevorotation". What is the difference between the prefixes "d- and D-" and "l- and L-"?

6) Draw the chair conformation of α-D-glucose. Identify the axial and equatorial bonds and the plane of symmetry. What do $^4C_1$ and $^1C_4$ notations mean? Check the structure of α-D-glucose you have just drawn. Is it $^4C_1$ or $^1C_4$?

7) Draw raffinose and identify the type of glycosidic bonds in the structure.

8) Draw D-mannose. Draw its aldonic, aldaric and uronic acids that occur after oxidation.

9) Draw xylitol and discuss the reaction that is used to form it.

10) Draw the structure of maltose and sucrose and identify their anomeric carbon. Explain why sucrose is not a reducing sugar.

11) Draw the structure of amygdalin, identify the type of bonds, and discuss its structure. Name three flavonoid glycosides.

12) **Online activity** Find online the structure of steviol glycosides and identify its aglycone.

13) **Online activity** Find online the terms "birefringence" and "Maltese cross" and their relationship to starch.

14) Draw the structure of chitin and identify the acetyl group, the amine group, and the type of glycosidic bond in the structure.

15) Draw and describe the structure of amylopectin. Identify the type of glycosidic bonds on the structure.

16) Draw the structure of hydroxy propylated starch and OSA starch. Identify the functional groups and the positions where the substitution occurs.

17) Draw and describe the starch pasting curve.

18) Draw and discuss the gelation mechanism of alginates and LM-pectin. Name two foods that use pectin gels.

19) Name the major monosaccharides of guar gum, gum Arabic and xanthan. Name two food applications of each.

20) What are hemicelluloses? Draw the structure of arabinoxylan and identify the anomeric carbons.

21) **Further reading**: Find the following book chapter "Tuvikene, R. (2021).
    Chapter 25 – Carrageenans. In G. O. Phillips & P. A. Williams (Eds.), Handbook
    of Hydrocolloids (Third Edition) (pp. 767-804): Woodhead Publishing".

    - Describe the gelation mechanism of κ-carrageenan.
    - What is the role of K^+ in κ-carrageenan gelation?
    - Describe the interactions of carrageenan with milk proteins.
    - Describe the interactions of carrageenan with locust bean gum.
    - Identify the functional properties of carrageenan in non-dairy and dairy foods.

2.6.4   Fill the Gaps

1) A carbon atom with _____ different groups attached to it is called chiral
   carbon and indicated as _____.

2) Isomers where one structure is the reflection of the other are called _____
   and are _____ of each other.

3) Monosaccharides carry aldehyde and alcohol groups they may react to form
   _____.

4) In the Haworth projection, the α-anomer of D-monosaccharides has the -OH
   group _____ the plane of the ring whereas β-anomers have the -OH
   group _____ the plane of the ring.

5) The two most common conformers of pyranoses are the _____and
   _____conformers.

6) Oxidation of the terminal -OH with the aldehyde group remaining unaffected
   results in _____ _____.

7) When monosaccharides oxidise, another compound is reduced. Because of this
   property, monosaccharides are also known as _____ _____.

8) Polysaccharides with only one type of glycosidic bond are _____, whereas
   when they have multiple types, they are _____.

9) Polysaccharide functionality falls into two main classes: structure_____
   or structure _____.

10) Large and lenticular starch granules are called _____. Small and spherical
    are called _____.

11) Modified starches are changed with a _____or _____method to improve
    their _____ properties.

12) Retrogradation that refers to the _____of _____and
    _____.

13) Pure glucose has dextrose equivalents of _____ whereas starch of _____.

14) _____ is commonly used in cocoa particle stabilisation in dairy desserts or ice creams.

15) Pectin consists of two major structural classes: _____and _____.

16) Pectins are divided into _____ _____with DE > 50%, and _____ _____ with DE < 50%.

17) Gum Arabic is commonly used to emulsify _____ oils.

18) Xanthan forms highly viscous solutions at very _____ concentrations.

19) The shell of lobsters and crabs is made of _____.

20) Arabinoxylan is a typical example of a _____.

Chapter 3
Proteins-Enzymes

Learning Objectives

After studying this chapter, you will be able to:

- Classify amino acids based on their chemical structure
- Describe the pH-dependency of amino acid and protein charge
- Classify food proteins based on their solubility and source
- Discuss the different levels of protein structure
- Discuss protein denaturation and hydrolysis
- Discuss the functional properties of proteins in food processing
- Describe the mechanism of enzymatic reactions
- Discuss the factors that influence enzymatic reactions
- Describe the action of major food enzymes on their substrates

3.1 Introduction

Proteins are macromolecules with numerous biological functionalities in living tissues. For example, they are involved in energy production, food digestion, or muscle contraction. From a nutritional point of view, they supply essential amino acids necessary to sustain life and the body's functionality. Food proteins must be extractable in high yields and suitable for further processing. For example, while potato protein is generally of high biological value, as it is rich in essential amino acids, the amounts found in potato are negligible, making it challenging to isolate large quantities of commercial interest. In contrast, whey proteins, also rich in essential amino acids, can be easily collected during cheese manufacturing and processed to appropriate form for further use (e.g., whey protein powder). Proteins used for food are of animal or plant origin. They can be consumed with minimal processing from their natural source (e.g., a glass of milk or a steak), extracted with

© The Author(s), under exclusive license to Springer Nature Switzerland AG 2021
V. Kontogiorgos, *Introduction to Food Chemistry*,
https://doi.org/10.1007/978-3-030-85642-7_3

suitable technologies and used as an ingredient in food formulations (e.g., whey protein in a cake mix), or processed into unique products (e.g., cheese, cold hams, or tofu). Animal protein sources are primarily meat, milk, eggs, and fish. Soy is the most common plant-based protein source, but new plant sources are explored, such as peas, beans, or lentils. The need for high-quality protein with a suitable nutritional profile and functionality has resulted in intense research interest to identify new protein sources such as those extracted from microorganisms (microbial protein), insects, algae, or even lab-grown protein (cultivated meat). A suitable nutritional profile is relatively easy to find in alternative protein sources. However, the factors limiting their broader use are poor (or unknown) technological functionality such as low water solubility, off-flavours, inadequate structure-formation properties, or problems with oxidative stability. In addition, their sensory profile and overall attributes do not always meet consumer expectations. For example, algae proteins have a fishy odour, and insect protein currently finds difficulties with consumer acceptability.

Food-protein science is distinct from classic protein biochemistry encountered in other fields of science (e.g., human or animal biology), as it is mainly concerned with the properties of proteins from a material science point of view. For example, a food scientist should manipulate and control structures created with proteins in foods (e.g., gels or emulsions) to enhance shelf life stability or improve textural properties. As a result, the tools and techniques needed to investigate protein functionality are not limited to those encountered in classic biochemistry (e.g., amino acid sequencing, structure characterisation, etc.). Still, they include techniques used in material sciences, for example, rheology, calorimetry, or particle size distribution analysis. Another notable difference is the *scale* and *purity* of the protein. A biochemist is usually concerned with a few milligrams of pure protein, which has been isolated under precisely controlled conditions. In contrast, a food-protein scientist deals with kilograms of protein at a time that operates in highly complex formulations that may simultaneously include sugars, lipids, and other compounds. Regardless of the different approaches, the principles of protein science derived from fundamental biochemistry are required to be well-understood before embarking on further exploration of protein functionality in the realm of material science.

3.2   Amino Acids

Amino acids are the building blocks of proteins. They consist of an amino group (–NH$_2$), a carboxyl group (–COOH), a hydrogen atom, and a *side chain* (R–) linked to a carbon atom called α-carbon (Fig. 3.1a). Amino acids are chiral when R– is not hydrogen, which occurs only in glycine. As a result, amino acids are found in enantiomeric forms D- and L- similar to carbohydrates, and we can draw them using a Fisher projection. In amino acids, the distinguishing characteristic in the Fisher projection is the positioning of the amino group when the –COOH group is placed at the top of the structure (Fig. 3.1b). If -NH$_2$ is located on the left of the structure, the

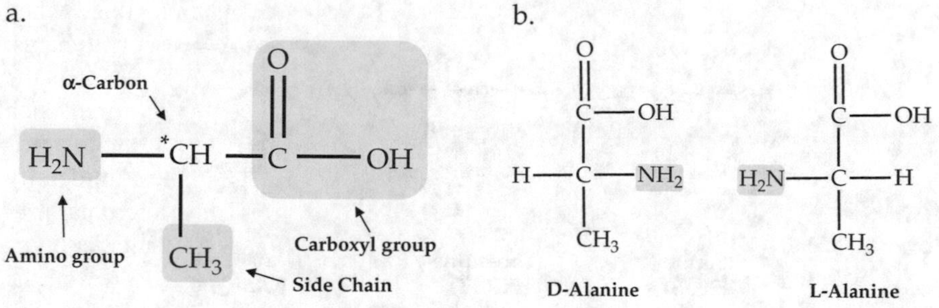

a.

b.

D-Alanine   L-Alanine

Fig. 3.1 (**a**) Structure of α-amino acids. α-Carbon has four different substituent groups and is chiral, and (**b**) Fischer projection of D- and L-alanine

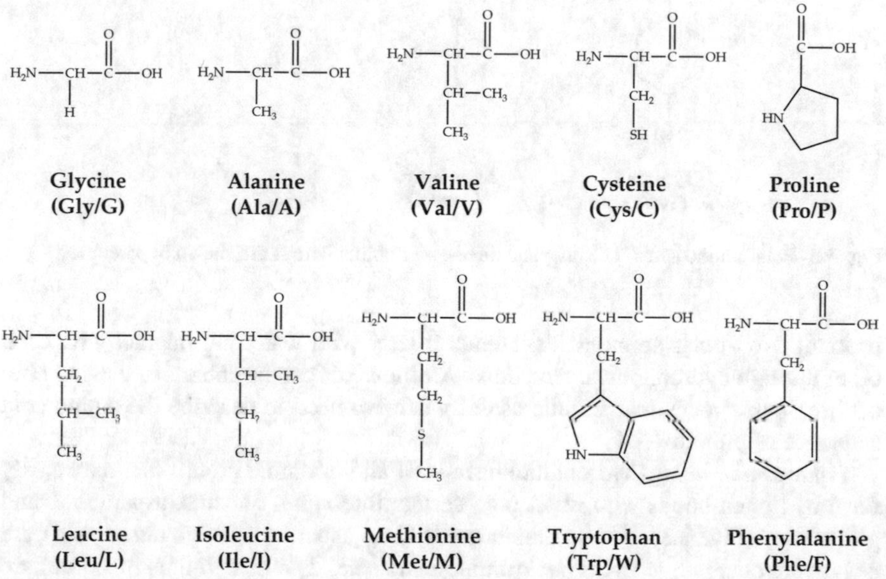

Glycine	Alanine	Valine	Cysteine	Proline
(Gly/G)	(Ala/A)	(Val/V)	(Cys/C)	(Pro/P)

Leucine	Isoleucine	Methionine	Tryptophan	Phenylalanine
(Leu/L)	(Ile/I)	(Met/M)	(Trp/W)	(Phe/F)

Fig. 3.2 Non-polar amino acids. The one- and three-letter abbreviations are shown in parentheses

amino acid is L- whereas if it is on the right, it is D-. Most natural amino acids are in L- configuration in contrast to natural sugars found in D-form.

The chemical properties of the side chain are the distinguishing characteristic of amino acids between each other. They are classified using different criteria depending on which property is important for the problem at hand. The most usual classification scheme is based on their capacity to interact with water resulting in four classes, i.e., *non-polar*, *polar*, *acidic*, and *basic*. *Non-polar amino acids* (hydrophobic) contain mostly hydrocarbon R groups and do not have either positive or negative charges (Fig. 3.2). Two types of hydrocarbon side chains are found in hydrophobic amino acids: *aromatic* (i.e., tryptophan and phenylalanine) and *aliphatic* (i.e., glycine, cysteine, methionine, alanine, valine, leucine, isoleucine, and

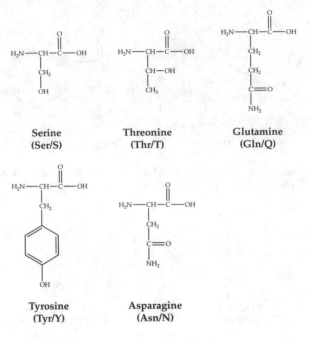

Fig. 3.3 Polar amino acids. The one- and three-letter abbreviations are shown in parentheses

proline). Non-polar amino acids interact poorly with water but interact with each other through hydrophobic interactions. Amino acids can be abbreviated using one- or three-letter codes that is quite useful when we need to describe the amino acid sequence of proteins.

Polar amino acids (hydrophilic) have -OH and/or amide groups that may easily form hydrogen bonds with water (i.e., serine, threonine, tyrosine, asparagine, and glutamine) (Fig. 3.3). *Acidic amino acids* (i.e., aspartic acid, glutamic acid) are negatively charged, whereas *basic amino acids* (i.e., lysine, arginine, histidine) are positively charged at pH 7.0 and may interact with each other through ionic interactions (Fig. 3.4). Amino acids may also be classified according to their nutritional significance into *essential, non-essential* (i.e., A, D, N, E, S) or *conditionally essential*. Nutritional classification may be necessary for some food formulations to create nutritionally balanced food. *Essential amino acids* are those that the body does not synthesise and must be obtained from the diet (i.e., H, I, L, K, M, F, T, W, V). *Conditionally essential* are those essential in the human diet under certain circumstances, e.g., disease or early development (i.e., R, C, Q, G, P, Y). A *limiting amino acid* is an essential amino acid found in insufficient quantities in a food source. *Most plant proteins have a limiting amino acid* e.g., lysine in grains, nuts, and seeds, threonine in grains, methionine in beans, and tryptophan in corn.

An important characteristic of amino acids is that their ionisation state changes depending on the pH. For that reason, they are called amphoteric, which means they

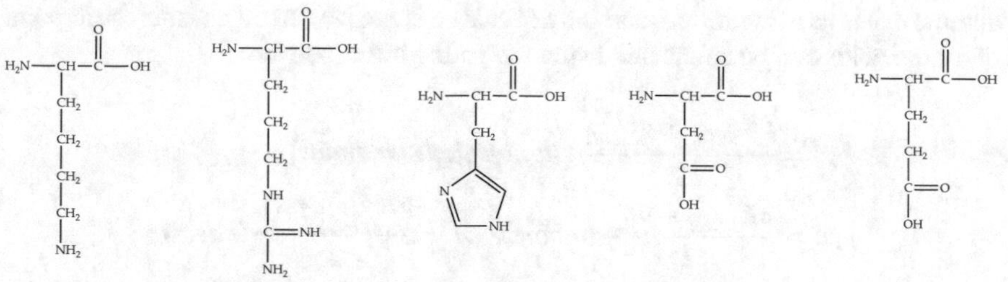
Lysine
(Lys/K)
Arginine
(Arg/R)
Histidine
(His/H)
Aspartic Acid
(Asp/D)
Glutamic Acid
(Glu/E)

Fig. 3.4 Positively (Lys, Arg, His) and negatively (Asp, Glu) charged amino acids at pH 7.0. The one- and three-letter abbreviations are shown in parentheses

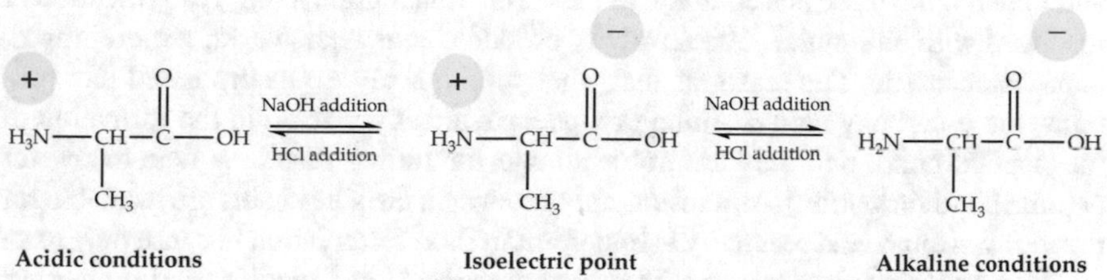
NaOH addition
HCl addition
NaOH addition
HCl addition
Acidic conditions
Isoelectric point
Alkaline conditions

Fig. 3.5 Amphoteric nature of amino acids. The charge depends on pH. The isoelectric point is the pH at which the net charge is zero

can react as acids or bases and form *zwitterions* (*dipolar ions*). At acidic conditions (low pH), the amino group is ionised, and amino acid carries a positive charge. In contrast, the carboxyl group is ionised at basic conditions (high pH), and the amino acid has a negative charge. The *isoelectric point* (pI) is the pH value at which amino acids have a net charge of zero, as they exist primarily as zwitterions (Fig. 3.5). The exact pH at which the transition occurs (from positively to zwitterion to negatively charged) depends on its structure and the pK_a values of each amino acid. Amino acids have up to three pK_a values: one for the α-carboxyl group (pK_{COOH}), one for the α-amino group ($pK_{NH3}{}^+$) and one for the side chain (pK_R). We can demonstrate the changes in amino acid ionisation using alanine with $pK_{COOH} = 2.4$ and $pK_{NH3}{}^+ = 9.7$ (Fig. 3.5). We start with alanine dissolved in a buffer with pH 1.0. At this pH, alanine is positively charged, as the pH is below pK_{COOH} and $pK_{NH3}{}^+$. We then add NaOH to increase pH to just above 2.4. At $pH > 2.4$ the hydrogen from the α-carboxyl group dissociates and the amino acid acquires a negative charge in addition to the positive charge that was already present in the structure (Fig. 3.5). We then keep adding NaOH until the pH gets just above 9.7. Above this pH, dissociation of hydrogen from the α-amino group results in loss of the positive charge and the amino acid acquires a net negative charge. At pH values between 2.4 and 9.7, alanine exists as a zwitterion, and the net charge is *nearly* zero. The isoelectric point of amino acids can be calculated using the following equations:

$$pI = \frac{pK_{COOH} + pK_{NH_3^+}}{2} \left(no\, charged\, side\, chain\right)$$

$$pI = \frac{pK_{COOH} + pK_R}{2}(acidic), \quad pI = \frac{pK_{NH_3^+} + pK_R}{2}(basic)$$

Using the pK_a values shown above and the first equation, the isoelectric point of alanine is pI = 6.0. The pK_{COOH} pK_{NH3}^+ and pK_R values of each amino acid can be easily found online. The relationship between pH and ionisation of amino acids is very important, as it also affects the properties of proteins and needs to be well understood before proceeding any further.

The α-carboxyl and α-amino groups of amino acids exhibit chemical reactivity and undergo the reactions common to these functional groups (e.g., esterification or reactions with ammonia). The reactivity of side chains is, however, a more important characteristic. The reason is that when amino acids are incorporated into proteins, the α-carboxyl and α-amino groups have already reacted in the formation of the peptide bond, and they are not available for further reactions (see below for peptide bond formation). As a result, only groups on the side chains are available for reactions. Amino acid reactivity is important in food formulation because they react during processing and storage with other components and modify food properties. Peptide bond *formation*, disulfide bond *formation*, and Schiff base *formation* are the most relevant reactions.

Amino acids are linked *via* a special bond called a *peptide bond*. This bond is formed between the α-carboxyl and α-amino groups of two amino acids (same or different) and proceeds *via* a condensation reaction (i.e., loss of water) (Fig. 3.6). Following peptide bond formation, one end of the newly formed molecule carries the -NH$_2$ group (*N-terminal*) and the other the -COOH group (*C-terminal*). By convention, N-terminal is drawn on the left and C-terminal on the right.

When two amino acids are linked through a peptide bond, the product is called a *dipeptide*, three a *tripeptide*, etc. As a guideline, when up to ~50 *amino acid residues* are linked, the chain formed is termed *polypeptide*, whereas when it consists of more than ~50, it results in *protein* formation. Peptides are commonly formed during food fermentations with desirable or undesirable effects on food flavour. In certain cases, bitter peptides may form (e.g., the tripeptide leucine-leucine-leucine), resulting in detrimental sensory attributes and product losses. Bioactive peptides may also be formed during fermentation or be naturally present in foods and may

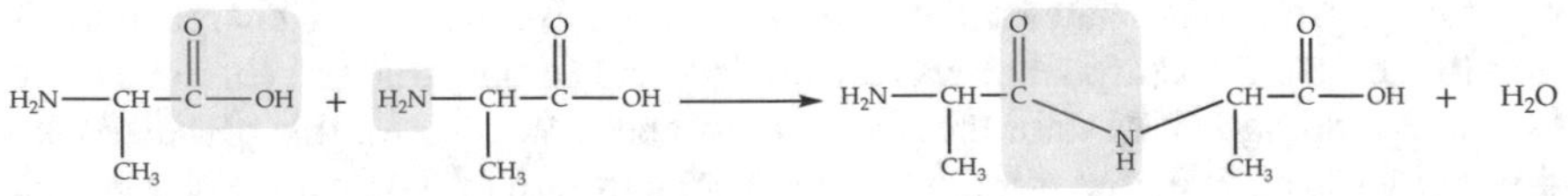

Fig. 3.6 Peptide bond formation between two amino acids

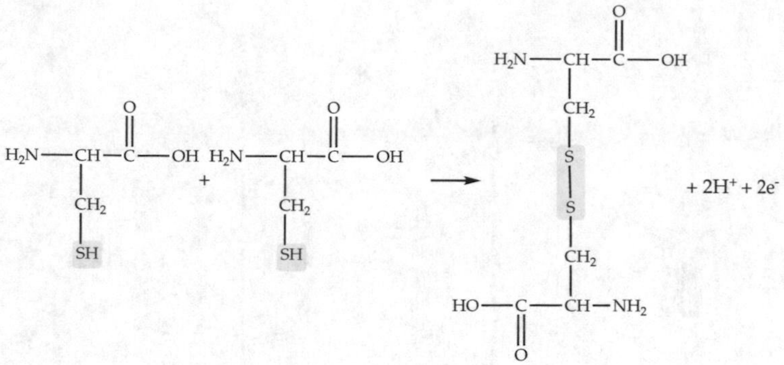

Fig. 3.7 Disulfide bond formation in cysteine. The resulting dipeptide is cystine. Notice the spelling difference

provide some form of health benefits (e.g., *casein hydrolysates*). Protein hydrolysates resulting in complex peptide composition are commonly used as flavour enhancers (see Chap. 8). Some peptides may have a sweet taste, such as the dipeptide *aspartame* and *neotame*, and are used as artificial sweeteners. Bacteriocins are produced by bacteria and are peptides or small proteins with antibiotic properties. *Nisin* is an approved bacteriocin that is frequently used as a food preservative.

The sulfhydryl group (-SH) of cysteine is highly reactive, and -SH groups of two cysteines oxidise easily to form cystine (Fig. 3.7). The bond between two -SH groups is called disulfide bond or disulfide bridge. It is important for the stability of proteins as well as for various technological applications of food ingredients. For example, the bread-making capacity of wheat flours is entirely controlled by the ability of gluten proteins to form disulfide bonds with each other.

Free amino groups of amino acids may react with aldehydes or ketones to form *Schiff bases* (*imines*) (Fig. 3.8). An imine is a chemical compound that contains a carbon-nitrogen double bond. Schiff bases are encountered very frequently in foods, as the presence of amino acid and a reducing sugar is almost guaranteed. They are important intermediate products of Maillard reactions resulting in compounds with intense colour and odour (see Chap. 6). In addition, the formation of lysine-Schiff bases during the Maillard reaction removes lysine from food and reduces its nutritional quality, as lysine is an essential amino acid. A final reaction worth mentioning is the reaction of amino acids with *ninhydrin* used to detect and quantify free amino acids.

3.3 Proteins

3.3.1 Protein Classification

Proteins have the most diverse functionality compared to the other molecules encountered in food and may be classified depending on their functionality. *Enzymes* are important in food processing and preservation and are discussed in the following

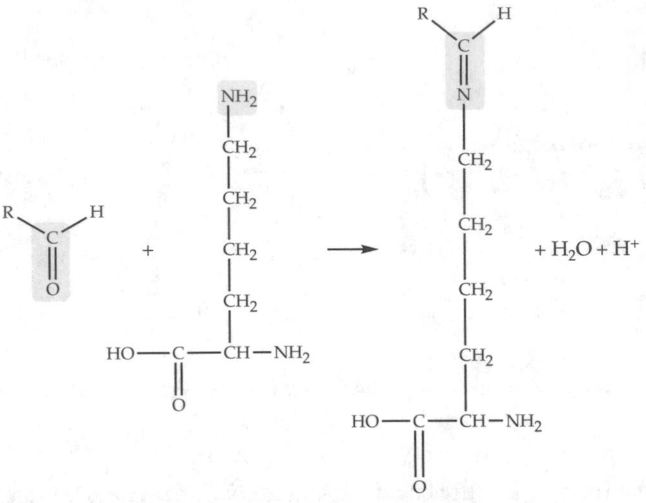

Fig. 3.8 Schiff base formation from lysine. Amino groups of amino acids may react with carbonyl groups to form compounds with a carbon-nitrogen double bond

section. *Structural proteins* are fibrous proteins such as keratin or collagen that are found in the skin or connective tissues of animals. Collagen, after appropriate processing, is converted to gelatin that is a commonly used food ingredient. *Contractile proteins* such as myosin or actin are found in muscle tissues and are responsible for muscle contraction. They are the main proteins ingested during the consumption of products of animal origin. For example, the generic term "steak" refers to a type of meat that is sliced across the muscle fibres of the carcass and it essentially consists of myofibrillar proteins (myosin and actin ~50%), sarcoplasmic proteins (enzymes and myoglobin ~30%) and connective tissues (e.g., collagen ~15%). Storage proteins (egg albumen, seed proteins, or milk proteins) include most proteins used in the food industry. *Protective proteins* (toxins, allergens) may be of significance for allergic individuals (e.g., soy or gluten peptides) or in case that they cause some form of disease (e.g., botulism). Finally, *hormones* (e.g., insulin or growth hormone), *transfer proteins* and *antibodies* are proteins responsible for various metabolic activities in the body but do not have any functionality in food formulation.

Proteins may also be classified depending on their *solubility*. Albumins are soluble in neutral salt-free water, whereas globulins in neutral salt solutions. Glutelins are soluble in dilute acid or alkaline solutions and *prolamins* are soluble in 50–90% ethanol. This classification, also known as the *Osborne protein classification* of storage proteins, is important because it allows fractionation and isolation of most food proteins. While most food proteins are generally soluble in aqueous solutions, *glutenin* and *gliadin*, the main protein fractions of gluten, are prime examples of food proteins insoluble in water. Instead, they hydrate and form a strong *gluten network* transformed into

Table 3.1 Names of proteins from major food sources. The proteins presented are essentially the proteins we ingest when consuming foods with simple (e.g., boiled egg or tofu) or complex (e.g., a cake) composition

Food source		Major proteins
Meat/fish	Muscle tissue	Myosin, actin
	Connective tissue	Collagen – Also processed to gelatin
Egg	White	Ovalbumin, ovotransferrin, ovomucoid, ovomucin, and lysozyme
	Yolk	Low-density lipoprotein (lipovitellin), high-density lipoprotein, livetin, and phosvitin
Milk	Caseins	α_{s1}-, α_{s2}-, β-, and κ-casein
	Whey	β-lactoglobulin, α-lactalbumin, bovine serum albumin and immunoglobulins
Gluten	Glutenin	High- and low- molecular weight subunits
	Gliadin	α-, γ- and ω-gliadins
Soy		β-conglycinin (7S), glycinin (11S)
Beans, peas, lentils, and other legumes		Legumin, vicilin
Nuts and insects		Not sufficiently characterised and classified

bakery products with appropriate processing. However, *glutenin* is soluble in dilute acids (i.e., it is a glutelin) whereas *gliadin* is soluble in ethanol (i.e., it is a prolamin).

Depending on the source, food proteins usually have a specific name associated with them, especially when they constitute a major component of that protein source (Table 3.1). For example, whey protein is not a single protein but a mixture of different proteins consisting of approximately 65% β-lactoglobulin, 25% α-lactalbumin, and 8% bovine serum albumin. Table 3.1 is not exhaustive, as there are other proteins that we may also consume or use as ingredients for specialised applications. However, for all practical purposes, these are most of the proteins that we consume daily. For example, when foods have a complex protein profile, like a cake made of wheat flour, eggs, and milk, we essentially consume gluten, egg, and milk proteins.

3.3.2   Protein Structure

In proteins, *four levels of structure* may be distinguished: *primary, secondary, tertiary,* and *quaternary*. The primary structure is the sequence of amino acids in the polypeptide chain (Fig. 3.9). Changing the 20 common amino acids order makes it possible to get more than 2×10^{18} combinations of different peptides with a unique structure and chemical properties. The specific sequence of amino acids in the polypeptide chain and their interactions with each other determine the secondary structure of the protein.

Secondary structure refers to the specific geometrical arrangement of the polypeptide chain along one axis and is *due to hydrogen bonding* between peptide bonds. Secondary structure results in the folding of the polypeptide chain leading to a

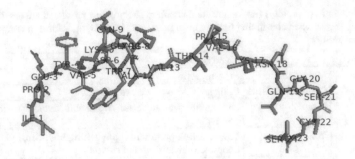

Fig. 3.9 A peptide with 24 amino acids with isoleucine being the first (ILE, number 1) and serine the last (SER, number 24). The sequence from 1 to 24 is the primary structure of this peptide. Changing the order of even one amino acid results in a different primary structure

Fig. 3.10 The α-helix. (a) Side view, (b) side view with the amino acid residues shown, (c) top view, and (d) close up view of one turn. A, I, S are alanine, isoleucine, and serine. Each turn (peak-to-peak) has 3.6 amino acid residues distinguished with different colours

unique *3D structure*. The driving force for this arrangement is to minimise exposure of hydrophobic sites of protein to water and maximise hydrophilic site exposure. Major types of secondary structures are *α-helices* and *β-pleated sheets*.

α-Helix is a coiled, rigid structure stabilised by hydrogen bonding parallel to the helix axis (Fig. 3.10). Hydrogen bonds form between the amino group of each amino acid and the carbonyl group of the amino acid four residues away. There are 3.6 amino acid residues per turn of the helix, and the *pitch*, the distance between corresponding points per turn, is 0.54 nm.

α-Helix has *amphiphilic nature,* with the one side of the helical surface being hydrophobic and the other hydrophilic. The overall hydrophobicity of a helix is determined by the amino acids present on the side chains. If the side chains are hydrophobic, the helix usually avoids water and is positioned within the protein structure, whereas if the side chains are hydrophilic, the helix is exposed to the water. The R- groups of amino acids extend outward from the helix (Fig. 3.10b). This conformation minimises the contact of hydrophobic parts of the helix with water while maximising the contact of the exterior parts, thus stabilising the structure. All amino acids participate in an α-helical structure except proline that interrupts its continuity.

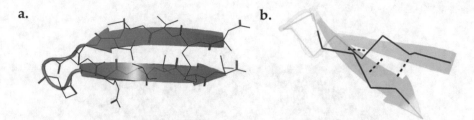

Fig. 3.11 Structure of β-pleated sheets. (**a**) Two β-strands with the amino acids shown in blue colour, and (**b**) A zig-zag line forms along the direction of the polypeptide backbone (red line). The two neighbouring β-strands interact with hydrogen bonds (black dotted lines). In this example, the three hydrogen bonds hold the structure together

Antiparallel β-strands Parallel β-strands Random coil

Fig. 3.12 Parallel and antiparallel β-strands and a random coil protein. A loop connects the β-strands

Polypeptide chains can also be found in an extended zigzag configuration known as *β-pleated sheets* consisting of *β-strands*. A β-strand is a stretch of polypeptide chain typically 3 to 10 amino acids long (Fig. 3.11a). β-Sheets are connected alongside each other to form the structure, in which the adjacent β-strands run in the same (*parallel*) or opposite (*antiparallel*) directions (Fig. 3.12). The adjacent protein chains are held together by hydrogen bonds (Fig. 3.11b).

Many proteins contain combinations of α-helices and β-sheets known as *motifs*. For example, a protein may have a *βαβ* unit: two parallel β-pleated sheets connected by an α-helix segment. *β-Turn* is a secondary structure in proteins where the polypeptide chain changes its direction (called a loop) and is regularly connected to strands of antiparallel β-sheets (Fig. 3.12). Proline and glycine are common amino acids found in β-turns. Other common motifs that may be observed are referred to as *β-meander*, *β-barrel*, *αα-unit*, or the *Greek key*. Finally, *random coils* do not have structure and the polypeptide chain is oriented randomly in space (Fig. 3.12). Two examples of the enzyme papain and egg-white protein are shown in Fig. 3.13 where different secondary structures combine to form the entire proteins.

The three-dimensional structure of proteins, in which the polypeptide chains are tightly folded and packed into a compact form, like that of papain shown in Fig. 3.13, is termed tertiary structure. There are two common types of tertiary structures:

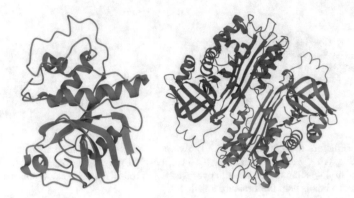

Fig. 3.13 Structures of papain found in papaya (left) and egg ovalbumin (right). Secondary structures of α-helices, β-sheets and random coils that form the proteins are easily distinguished. Figures were drawn in PyMOL based on PDB accession codes 9PAP and 1OVA

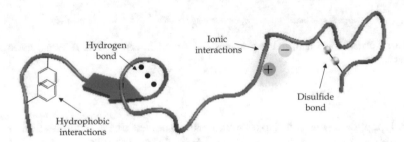

Fig. 3.14 Forces that stabilise the tertiary structure of proteins

fibrous and *globular*. Fibrous proteins are rod-like structures and are relatively heat and pH-stable whereas globular proteins form spherical structures. Tertiary structure is stabilised by *hydrophobic interactions* where hydrophobic amino acids interact to avoid contact with water, *electrostatic interactions* (*salt bridges*) between positively charged and negatively charged groups, *hydrogen interactions*, and *disulfide bonds* when two cysteines react with each other (Fig. 3.14).

Finally, proteins with more than one chain may have a quaternary structure that is the number and arrangement of multiple protein subunits in a complex (Fig. 3.15). These proteins are composed of *oligomers* that are *macromolecular complexes* composed of non-covalent assemblies of two or more *monomer subunits*. Monomer subunits interact with hydrophobic interactions to form dimers, trimers, or tetramers for two, three or four monomers. Example of such proteins are β-conglycinin and glycinin from soybeans or β-lactoglobulin from milk.

Fig. 3.15 Quaternary structures of β-conglycinin from soy (left) and phaseolin (right) from beans forming trimers (green, brown, and blue). Structures seem to be similar, as they are both vicilins, the storage proteins of soy or beans. Examination of their primary structures can reveal their differences. Figures were drawn in PyMOL based on PDB accession codes 1UIK and 2PHL

3.3.3 Changes in Protein Structure: Denaturation and Hydrolysis

Proteins may lose their structure during processing, storage, or food digestion. Denaturation is the loss of secondary, tertiary, or quaternary structure of proteins, resulting in *protein unfolding* (Fig. 3.16). Denaturation **does not involve peptide bond cleavage** and frequently, but not always, is reversible.

Denaturation affects protein structure and functionality in many ways. Coagulation or gelation is the random aggregation of denatured protein molecules due to protein-protein interactions most frequently due to high temperatures or changes in pH. As protein unfolds, hydrophobic amino acids buried in the structure are now exposed and can interact through hydrophobic interactions, which results in gel *formation*. Common examples include the coagulation of milk proteins during yoghurt manufacturing when pH is lowered and heat-induced denaturation of egg proteins. At this point, it should be emphasised that changes in temperature (heat) and pH are the two most common denaturation methods in food processing. Denaturation also results in *loss of biological activity*, which is important for enzymes. Inactivation of enzymes by heat treatment is a common processing step to increase the shelf life of various food products. For instance, *blanching* is a low-temperature heat treatment to denature enzymes naturally present in vegetables before freezing. If left uncontrolled, these enzymes have detrimental impact on the product's quality during sub-zero temperature storage. Enzymes in the gastrointestinal tract digest easily denatured proteins. As a result, denaturation *increases the bioavailability* of amino acids and other compounds associated with that protein (e.g., iron). High-temperature processing of certain foods (e.g., canned tuna) denatures proteins and makes them digestible, conferring desirable sensory characteristics and microbiological stability. Finally, unfolding and exposure of hydrophobic

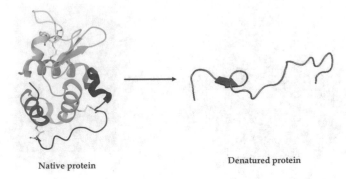

Native protein   Denatured protein

Fig. 3.16 During denaturation, proteins unfold without peptide bond cleavage

sites improve the *emulsification and foaming capacity* of proteins, making them sometimes irreplaceable in certain food formulations (e.g., salad dressings).

Protein Gelation

Proteins, similar to polysaccharides, form gels under the appropriate conditions. Gelation is an important property of proteins, as they are one of the main ingredients responsible for food structuring, i.e., holding things together. Without protein gelation, most of our food would resemble a thick custard. The most common method to form food gels involves heating. Heat causes protein denaturation that exposes reactive groups that favour intermolecular interactions between protein chains. Once more, the involved forces are the same as in polysaccharide gels (i.e., hydrogen, ionic, and hydrophobic interactions). Hydrophobic interactions are more important in protein than in polysaccharide gels due to hydrophobic amino acids present in most proteins. Protein gels are either *fine-stranded* or *particulate*. Particulate gels (e.g., yoghurt) are opaque and have a low water-holding capacity, whereas fine-stranded gels (e.g., gelatin) are transparent or translucent and have a high water-holding ability.

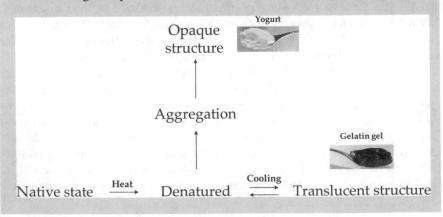

Table 3.2 Agents that influence protein denaturation

	Agent	Comments
Physical	Heat	Heat processing
	Mechanical shear	Mixing, extrusion, homogenisation, high shear processing
	Pressure	Reversible with pressure removal
	Ionising radiation	Processing by irradiation
	a_w	Low a_w enhances thermal denaturation stability
Chemical	Organic solvents	Weaken hydrophobic interactions
	pH	Proteins unfold at alkaline pH, precipitate at pI. Influences electrostatic interactions.
	Urea, guanidine hydrochloride	Weaken hydrogen bonds
	Mercaptoethanol and dithiothreitol	Break disulfide bonds
	Detergents (e.g., SDS)	Weaken hydrophobic interactions
	Sugars	Enhance thermal denaturation stability
	Salts	See Hofmeister series

Denaturation is caused by *physical* or *chemical agents* (Table 3.2). *Physical agents* include *heat, mechanical shear, pressure, irradiation,* and *water activity. Heat,* the most common process used in the food industry, causes protein denaturation, as vibrations at high temperatures break protein intramolecular interactions (does not break peptide bonds). Amino acid composition affects thermal stability and heat-stable proteins usually contain a high proportion of hydrophobic amino acids. Water facilitates thermal denaturation and proteins in high moisture foods can be easily denatured, in contrast to those found in dry powders. Generally, low water activity enhances thermal denaturation stability of proteins. *Mechanical shear* refers to agitation such as whipping, shaking, kneading, or mixing encountered in several food processing operations. The input of mechanical energy destabilises protein structure leading to the stretching of protein molecule and denaturation. Denaturation due to whipping is important in foam formation used to create unique textures in various products (e.g., meringues). *Hydrostatic pressure* and *irradiation* used in non-thermal (athermal) processing of foods also cause denaturation by breaking the forces that stabilize the tertiary structure. Pressure is important in high hydrostatic pressure processing (HHPP), where pressure is elevated to 1–12 kbar. Denaturation during HHPP occurs due to compressibility of protein structure and is mostly reversible with removal of pressure. Depending on the dose, *ionising irradiation* may cause unfolding, coagulation, and even cleavage of peptide bonds. Its effect on food proteins is quite complex and it depends on the specific food that is being irradiated and the objectives of the treatment (i.e., dose). *pH* plays a critical role in protein properties. Proteins unfold at extreme pH values, and unfolding is greater at alkaline pH than at harsh acidic environments. pH-induced denaturation is frequently reversible if there is no peptide bond cleavage. Changes in pH influences protein charge similar to amino acid ionisation, as it affects electrostatic interactions between charged groups that belong to the protein. Proteins are most stable at their isoelectric point, where the net charge is zero. The presence of *organic solvents* such as ethanol affects mainly the stability of hydrophobic interactions. Hydrophobic interactions are weakened, and protein unfolds and precipitates. Electrostatic interactions are generally enhanced, but the overall effect depends on the protein-solvent pair. Ethanol is the only edible organic solvent used in food processing or produced during the fermentation of various products. Ethanol induces protein denaturation, altering the product's stability and sensory characteristics. In certain alcoholic beverages (e.g., Irish creams that are alcoholic beverages made with cream, cocoa, and whiskey), the coexistence of protein and ethanol makes the formulation of such products challenging. Although ethanol is the only edible solvent other solvents may also be used as processing aids that may influence protein functionality to a certain extent (e.g., hexane use in soybean processing operations). *Low molecular weight molecules* with the capacity to influence molecular interactions may be present in the food or used in a laboratory setting in protein-related investigations. For example, *urea* or *guanidine hydrochloride* weaken the strength of hydrogen bonding, causing denaturation that may be reversible after denaturant removal. *Mercaptoethanol* and *dithiothreitol* both break disulfide bridges destabilising the tertiary protein structure. *Detergents* such as sodium dodecyl sulfate (SDS) are powerful denaturing agents that disrupt hydrophobic interactions, causing protein unfolding. *Sugars* (e.g., sucrose) tend to stabilise the proteins, whereas *salts* have a complex effect on protein denaturation. *Salts* may cause denaturation, as they compete with protein for interactions with water molecules. Anions influence protein structure more than cations. At sufficiently high salt concentrations, few water molecules are available to interact with protein to hydrate it and as a result protein aggregates and precipitates. Salt-induced protein precipitation is called salting out, whereas protein solubilisation, because of appropriate amounts of salt, is termed salting in. Not all ions have the same capacity to denature proteins. The Hofmeister series or *lyotropic series of salts* is a classification of salts in order of their ability to precipitate (or solubilise) proteins (Fig. 3.17). The most commonly used salt in foods is NaCl, which is in the middle of the series and has intermediate aggregation (or solubilisation) properties. Other commonly used inorganic salts are $NaNO_3$, $CaCl_2$, or phosphates while organic salts may also be used (citrates, acetates, benzoates, or sorbates). A common salt used for research purposes is ammonium sulfate (($NH_4)_2SO_4$). With both ionic species (NH_4^+ and SO_4^{2-}) being at the left of the scale,

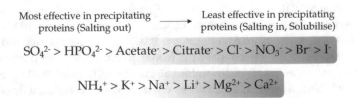

Most effective in precipitating ⟶ Least effective in precipitating
proteins (Salting out)   proteins (Salting in, Solubilise)

$$SO_4^{2-} > HPO_4^{2-} > Acetate^- > Citrate^- > Cl^- > NO_3^- > Br^- > I^-$$

$$NH_4^+ > K^+ > Na^+ > Li^+ > Mg^{2+} > Ca^{2+}$$

Fig. 3.17   Hofmeister series of anions and cations relevant to food proteins

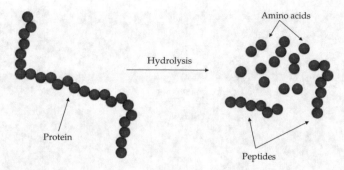

Fig. 3.18 Protein hydrolysis results in cleavage of the peptide bond and formation of peptides and free amino acids

it exhibits a very strong capacity to precipitate proteins and is used in protein isolation processes.

In contrast to denaturation, where the breaking of peptide bonds does not occur, *hydrolysis* involves cleavage of peptide bonds yielding smaller polypeptides or free amino acids (Fig. 3.18). It is achieved by the action of acid and heat (*acid hydrolysis*) or proteolytic enzymes (*enzymatic hydrolysis*). The main difference between these two modes of action is that acid hydrolysis is a random process cleaving the protein uncontrollably. Hydrolysis products depend on the conditions (e.g., heat, time, type of acid, etc.), and changing conditions most certainly changes the end products. In contrast, enzymatic hydrolysis cleaves only one bond, always resulting in the same products. Both hydrolysis types are useful depending on the objective of the process. For instance, protein hydrolysates used as flavourings may be produced with acid hydrolysis resulting in a complex mixture of peptides and free amino acids with a strong savoury taste. In contrast, rennin used in cheese manufacturing breaks only one bond.

Protein Determination

The internationally accepted method for protein determination in foods is the *Kjeldahl method*. The method does not measure protein but the total organic nitrogen released from a sample and absorbed by acid. The Kjeldahl method has three steps. In the *digestion step*, the sample reacts with concentrated sulfuric acid, and the nitrogen released is converted to ammonium sulfate. *Distillation* involves adding excess base to ammonium sulfate to convert NH_4^+ to NH_3 that is distilled into a boric acid solution. *Titration* involves quantifying the amount of ammonia in the receiving solution. A conversion factor is finally used to convert total nitrogen to protein content.

Digestion

$$Protein \xrightarrow[\text{Heat, catalyst}]{\text{Sulfuric acid}} \underset{\text{ammonium sulfate}}{(NH_4)_2SO_4} + H_2O + CO_2$$

Distillation

$$(NH_4)_2SO_4 + 2NaOH \longrightarrow 2NH_3 + Na_2SO_4 + 2H_2O$$

$$\underset{\text{ammonia}}{NH_3} + \underset{\text{boric acid}}{H_3BO_3} \longrightarrow \underset{\text{ammonium-borate complex}}{NH_4^+ : H_2BO_3^-} + \underset{\text{excess boric acid}}{H_3BO_3}$$

Colour change

Titration

$$2NH_4^+ : H_2BO_3^- + H_2SO_4 \rightarrow (NH_4)_2SO_4 + H_3BO_3$$

Reverse colour change

When pure proteins are isolated, other methods are usually used to determine protein depending on the needs and purpose of the analysis. Examples include the Biuret method, Lowry assay, bicinchoninic acid (BCA) assay, absorption at 280 nm or Bradford assay, all with their advantages and disadvantages.

3.3.4 Functional Properties of Proteins

Except for their nutritional value, proteins have very important functional roles in food manufacturing. Most of their functional properties are similar to polysaccharides, and they are frequently exploited simultaneously (Table 3.3). For example, proteins are also able to gel and hold water within the structure. However, depending on the specific food, we may need both in the formulation but for different purposes. For example, the structure of crème caramel (a dairy dessert) is formed because of milk protein gelation. Milk proteins are responsible for holding water and form the structure in this product. However, prolonged storage leads to the collapse of the structure and expulsion of water (i.e., syneresis) accompanied by changes in the product's texture and aesthetics, limiting its shelf life. The addition of small amounts of polysaccharides helps retain water within the structure and prolong the shelf life. In this example, proteins are responsible for structure formation, water retention, and flavour binding, whereas the polysaccharide is responsible for water retention and texture modification. Proteins may also *modify viscosity*, but polysaccharides are mainly used as viscosity modifiers in most food applications.

Table 3.3 Functionality of proteins in foods

Property	Examples
Gelation	Boiled egg, yoghurt
Emulsification	Salad dressings
Foaming	Meringue, beer
Flavour binding	Retention of food flavours
Water retention	Water loss prevention from the structure
Enzymatic action	See next section

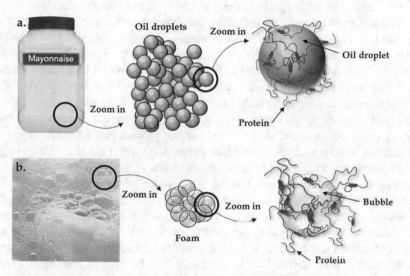

Fig. 3.19 (a) Emulsification, and (b) foaming of proteins. Proteins surround the oil droplet or bubble to form a protective film. The physicochemical properties of the film are crucial for the stability of both emulsions and foams

Proteins can form *foams* and *emulsions*, as they have both hydrophobic and hydrophilic parts in their structure. A denaturation step is most frequently required, as it results in unfolding that exposes the hydrophobic amino acids towards the hydrophobic phase, allowing the protein to adsorb. In emulsions, proteins create a protective thin film around the oil droplets and keep them dispersed. In foams, they arrange at the air-water interface and create a bubble entrapping air (Fig. 3.19). Emulsion and foam stabilisation achieved with proteins is critical in many foods. For example, the quality and sensory properties of salad dressings and sauces depend on emulsion characteristics. Meringues and many confectionery products rely on foam film stability. Ice cream needs both suitable emulsion and foam stabilisation during processing. In solid foams (i.e., sponge cakes), proteins stabilise air cell structure essential for volume and texture of the products (e.g., egg proteins in cake products). As is evident, the emulsification and foaming capacity of proteins are principal technological properties and optimisation of their performance is an active field of research. Finally, proteins also can *bind flavours*, thus improving or modifying the sensory properties of products (see Chap. 8).

3.4   Enzymes

Enzymes are proteins with *catalytic activity* (i.e., *catalysts*), meaning they speed up chemical reactions. Although enzymes accelerate the reaction they catalyse, they do not take part or consumed during the reaction. They are still in the same form when the reaction is complete and ready to participate in a new reaction cycle. Enzymes facilitate the course of a reaction that sometimes would be impossible without their presence. They are naturally found in foods and are important in the food industry since they have both desirable and undesirable influences on food quality. When natural enzymatic activity shortens shelf life, we may need to inactivate them to preserve the quality (e.g., browning of fruits and vegetables). On the other hand, enzymes may be important to confer the desired product characteristics, as in fermented products (e.g., cheese, salami, or pickles). Enzymes are also used as indicators of proper processing. For instance, deactivation of alkaline phosphatase in milk indicates correct pasteurisation, whereas deactivation of peroxidase and catalase in blanching before freezing ensures that the frozen product has optimum quality during storage. Similarly, lipoxygenase deactivation is also used as an *index enzyme* in vegetable processing (e.g., corn, peas, and green beans). Enzymes may also be used directly to manufacture a specific food ingredient, for example, amylases in glucose manufacturing from starch. Finally, enzymes may be used as analytical tools to measure another compound in a food product, e.g., measuring glucose using glucose oxidase.

3.4.1   Mechanism of Enzymatic Reactions

Molecules need to meet each other (collide) for chemical reactions to occur. Even when molecules collide, this needs to happen successfully to create products, and reactants must overcome some *energy barrier* for a *successful collision*. The energy that molecules must overcome for a chemical reaction to occur is called activation energy. If the barrier is too high, then the reaction is difficult, meaning more molecular collisions are needed. Enzymes decrease the activation energy that is required, thus facilitating the reaction (Fig. 3.20).

Enzymes act upon molecules that are called *substrates* and are so specific that they usually catalyse only one type of reaction. When a substrate is chemically different from what the enzyme requires, the reaction does not occur, but some enzymes may act upon many substrates (see *specificity* below). Enzymes with the same or similar enzymatic activities but different protein structures are called *isoenzymes*.

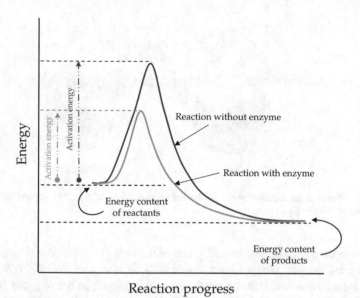

Fig. 3.20 Activation energy of enzymatic reactions. Enzymes reduce the activation energy needed for successful collisions to occur, making the reaction easier

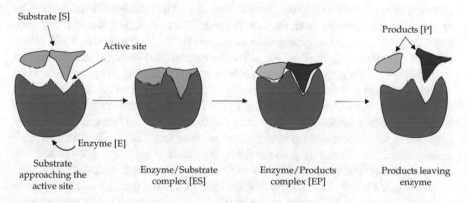

Fig. 3.21 Mechanism of enzymatic reactions. See text for description

Enzyme-catalysed reactions occur at a particular location on enzyme's surface called the active site (Fig. 3.21). The specific substrate can form relatively weak, temporary bonds to the enzyme at the active site. The specificity of the substrate-enzyme interaction can be better understood with the *lock-and-key analogy*. The substrate is represented with a key, whereas the lock is the enzyme. When the fit of the key to the lock is correct, then the door opens, i.e., a reaction occurs. When the key does not fit the lock, the door does not open, i.e., the reaction is not possible. In fact, the substrate often does not fit exactly on the active site. However, when the

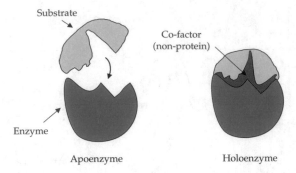

Fig. 3.22 When the substrate does not fit the active site, co-factors act as adapters to allow the attachment of substrates

correct substrate approaches the active site of the enzyme, it induces changes in the enzyme conformation (shape) for a closer fit. This is a more realistic depiction of how enzymes accommodate the substrates and is known as the *induced fit theory* (Fig. 3.21). Generally, speaking an enzymatic reaction occurs in four steps. In the first step, the substrate (denoted as [S]) approaches and enters the active site of the enzyme ([E]). Following this, an *enzyme-substrate complex* ([ES]) is formed. It is termed "complex", as a new temporary molecule is formed that is stabilised with transient interactions (e.g., hydrogen bonding etc.). The reaction occurs during the residence of the substrate in the active site, and after completion, the new complex that is formed is termed *enzyme-product complex* ([EP]), as the products still reside within the active site. In the final stage, the *product(s)* ([P]) leave the active site, and the enzyme is ready to participate in a new reaction cycle (Fig. 3.21).

Enzymes may be composed exclusively of protein (single or multiple subunits) or may have a more complex structure. In the latter case, a non-protein compound is usually required for the enzyme to carry out the reaction and is termed a co-factor (Fig. 3.22). Co-factors are divided into *metal ions* (e.g., Zn^{2+}, Cu^{2+}, or Mn^{2+}) or *small organic molecules* (e.g., vitamin-B complex). The latter, also called *co-enzymes*, are primarily vitamins and other organic essential nutrients (e.g., co-enzyme-Q_{10}). Co-enzymes are further divided into *prosthetic groups*, where the co-enzyme is covalently bound to the protein and *co-substrates* that interact with the active site with transient interactions (e.g., hydrogen or hydrophobic interactions) and is not permanently bonded to the enzyme. In the absence of the co-factor, the enzyme is inactive and is called apo-enzyme, while when the co-factor is in place, it is called holo-enzyme (Fig. 3.22).

Enzymes requiring metal in their structure for activity are known as *metalloenzymes*. Metalloenzymes bind and retain their metal atom(s) under all conditions with very high affinity. Those with a lower affinity for metal ions are known as metal-activated enzymes. Many food enzymes need metals to exert their catalytic activity, and a common strategy, if we need to inactivate them, is by removing the metal from the structure. For example, polyphenol oxidase (see Chap. 5) needs copper, alkaline phosphatase requires zinc, whereas and α-amylase needs calcium.

Two key terms associated with enzymatic function is specificity and selectivity. Specificity is the ability of an enzyme to catalyse only one reaction. For example, a lipase hydrolyses lipids but cannot hydrolyse proteins. Specificity arises from the 3D structure of the enzyme active site, as the substrate fits correctly and is optimally aligned to react. Selectivity is the preference of an enzyme towards competing substrates with similar structures. For example, the isomerisation of glucose into fructose can be catalysed by xylose isomerase. However, the natural reaction for xylose isomerase is the isomerisation of D-xylose into D-xylulose. Still, the activity towards glucose is sufficient for the industrial production of high-fructose corn syrups.

3.4.2 Enzyme Kinetics

Enzyme kinetics is the study of the rate at which an enzyme converts the substrate to products. It is essential to know the speed of the reaction because it allows us to use enzymes in industrially important applications efficiently and profitably. If the enzymatic reaction is too slow or the yields of converting the substrate to a product are low under the required processing conditions, the enzyme may not be suitable and needs to be replaced. In addition, using enzyme kinetics, we may study the factors that influence the reaction and find out how to accelerate or stop it. Factors that control reaction rate include all the factors that influence protein tertiary structure (see Sect. 3.3.3), enzyme type, enzyme concentration, substrate type, substrate concentration, presence of inhibitors, and water activity. As a general principle, any factor that changes the enzyme structure in an undesirable way (e.g., by denaturation or hydrolysis) reduces the speed or eliminates the enzymatic activity.

3.4.2.1 Effect of Substrate Concentration

The term used in enzymology to describe the speed at which a reaction occurs is *reaction velocity* (V), with units being concentration per unit time, e.g., mmol/min. It is important to know the *initial velocity* (V_o) and the *maximum velocity* (V_{max}) in enzymatic reactions. Initial velocity is the reaction rate at the beginning of the reaction when the enzyme and substrate have just been mixed. It is called "initial" because the enzyme starts using up the substrate, and the speed decreases and eventually drops to zero as there is no more substrate to catalyse. The relationship between V_o and [S] at low substrate concentrations is linear, i.e., increasing [S] results in a proportional increase of V_o (if we double [S], we double V_o, if we triple [S], we triple V_o etc.) (Fig. 3.23). This occurs because the active site of the enzyme is not *saturated*. As a result, the substrate can easily find a free active site and be converted to a product. With a gradual increase of [S], the active sites eventually become saturated (Fig. 3.23, yellow circle). Once the enzyme is saturated, the speed stays constant with further increments in [S]. This occurs because the substrate does

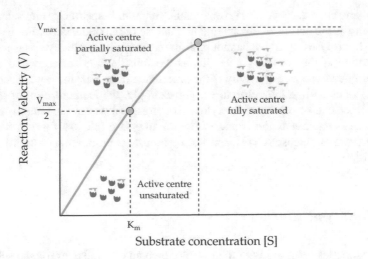

Fig. 3.23 Effect of substrate concentration on reaction velocity. This plot is known as the *saturation curve*

not have available active sites to be converted to a product. Consequently, it has "to wait" until a reaction cycle has been completed and an active site becomes available again (Fig. 3.23).

The velocity of the reaction at which all active sites are saturated with substrate and V_o is independent of [S] is called *maximum velocity* (V_{max}). V_{max} is the fastest velocity at which a given amount of enzyme can operate. Michaelis-Menten equation describes the relationship between V_o and [S]:

$$V = \frac{V_{max}[S]}{K_m + [S]}$$

with K_m being the Michaelis-Menten constant. Michaelis-Menten equation says that rection velocity V at any instant depends on K_m, V_{max}, and [S] of that moment. K_m is the substrate concentration that gives half of the maximum velocity ($V_{max}/2$) and is a measure of how fast V increases with substrate concentration (unit of K_m is molarity (M)) (green circle, Fig. 3.23). K_m is also a measure of the tendency of the enzyme to bind to its substrate (enzyme affinity). A low K_m corresponds to a high affinity for the substrate, i.e., the enzyme "likes" the substrate and requires a low substrate concentration to achieve V_{max}. Higher K_m corresponds to a lower affinity for the substrate, i.e., the enzyme "dislikes" the substrate and requires a high substrate concentration to achieve V_{max}. If an enzyme catalyses a reaction with two similar substrates (e.g., glucose and mannose), it prefers the substrate for which it has a lower K_m value. It becomes evident that by using K_m, we may decide if the enzyme is suitable for a particular reaction we need to accomplish under certain processing conditions.

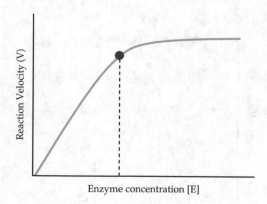

Fig. 3.24 Effect of enzyme concentration on the velocity of enzymatic reactions

Before non-linear curve-fitting on modern software was possible, it was difficult to estimate K_m and V_{max} from saturation curves, and scientists carried out various linear transformations of the Michaelis-Menten equation, using the Lineweaver-Burk, the Eadie-Hofstee or the Hanes–Woolf plots. Linear regression is used in all these methods to estimate the slope and intercept to calculate K_m and V_{max}. Although these methods are useful to visualise data, they are now largely obsolete. In modern biology, the analysis focuses on calculating these parameters using non-linear regression with computer software, as this technique allows direct determination of K_m and V_{max} from untransformed data points.

3.4.2.2   Effect of Enzyme Concentration

When the substrate is in excess (i.e., plenty of it), reaction velocity is proportional to enzyme concentration up to a certain point (purple circle, Fig. 3.24). At concentrations higher than those corresponding to the purple circle adding more enzyme does not influence velocity. The underlying principles of this behaviour are the same as those described in the previous section on the effect of substrate concentration.

3.4.2.3   Effect of Temperature, pH and a_w

Enzymatic reactions speed up as temperature increases. However, each enzyme has an *optimum temperature* at which it works best (Fig. 3.25). Below the optimum temperature, enzymatic activity slows down, whereas the enzyme starts to denature above the optimum temperature. *When the enzyme denatures completely, it loses its activity.* It should be emphasised that thermal denaturation of naturally occurring enzymes is the first tool (e.g., blanching) that a food technologist uses to control enzymatic activity before considering any other approach (e.g., changes in pH or

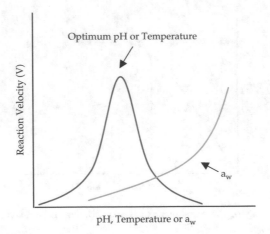

Fig. 3.25 Effect of pH, temperature, and water activity on enzymatic reactions

additives). The optimum temperature depends on the type of enzyme and the reaction it catalyses. For all practical purposes, it is usually around 45 °C, whereas beyond 50 °C the enzyme starts to denature. A class of enzymes, known as *heat-stable enzymes*, retain their activity at higher temperatures creating problems for the stability of some foods (see peroxidases below) or may be used for specialised applications as, for example, the heat-stable α-amylase that retains activity at temperatures >90 °C.

Similarly, enzymes function best within a specific pH range. Outside this optimum range, enzymatic activity decreases. The pH value where the enzyme is most active is known as the *optimum pH* (Fig. 3.25), and it varies depending on the enzyme and the reaction it catalyses. For example, pepsin, which works in the stomach, functions best in strongly acidic environments, whereas lipase found in the small intestine works best in alkaline environments.

Lowering *water activity* decreases enzymatic reaction rates and is usually eliminated below ~0.3, as the enzymes cannot function (Fig. 3.25). Dry products and powders with a_w < 0.2 are stable against enzymatic reactions. IMFs with a_w in the range of 0.3–0.8 require low-temperature preservation to minimise any undesirable enzymatic activity, especially when they are stored for extensive periods (6 months or more). Products with a_w higher than 0.8 require deactivation of enzymes with heating or other means to extend their shelf life.

3.4.2.4   Enzyme Inhibitors

Enzyme inhibitors are substances that alter the catalytic action of enzymes and consequently slow down, or in some cases, stop catalysis. Inhibitors bind to the enzyme either *reversibly* or *irreversibly*. The irreversible inhibitors (poisons) change the

Fig. 3.26 In competitive inhibition, the inhibitor binds to the active centre, and the substrate cannot fit. In non-competitive inhibition, the active site changes shape, and the substrate does not fit

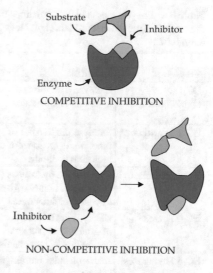

tertiary structure of the enzyme permanently and the enzyme cannot function again. Reversible inhibitors do not bind permanently and can be removed from the enzyme. Two major types of reversible inhibition can be distinguished: *competitive* and *non-competitive* (Fig. 3.26). Competitive inhibition occurs when the substrate and the inhibitor compete against each other for the active site of the enzyme. *Non-competitive inhibitors* are considered substances that, when added to the enzyme, alter the structure of the active site so that it cannot accept the substrate. Many enzymes contain sulfhydryl (-SH), hydroxyl (-OH), or carboxyl groups (-COOH) as part of their active sites. Any chemical which can react with these groups acts as a non-competitive inhibitor. Examples include heavy metals, such as Hg^{2+} or Pb^{2+}, which have strong affinities for -SH groups. The non-competitive inhibitor binds to the enzyme near to or away from the active site. If it is away from the active site, it is known as allosteric inhibition.

3.4.3 Enzyme Nomenclature and Classification

Enzymes are named by adding a suffix "*-ase*" to the name of their major substrate. For example, protease catalyses the hydrolysis of a protein, sucrase catalyses the hydrolysis of sucrose, whereas lactase catalyses the hydrolysis of lactose. There are exceptions to this rule, for example, pepsin, trypsin, or chymotrypsin. *Food enzymes* may be classified into *endogenous* when they are naturally present within food or *exogenous* when added for a particular technological purpose. Endogenous enzymes are responsible for various desirable or undesirable changes in raw and processed products, such as colour changes (e.g., polyphenol oxidase), flavour (e.g., lipase, lipoxygenase), or texture modification (e.g., proteases, amylases, or

Table 3.4 Enzyme classification, reactions they catalyse and importance for foods. Their relative importance is indicated with stars (1 star low, 2 stars intermediate, 3 stars high importance), and X: unimportant

Class	Reaction	Examples	Importance for foods
EC 1 Oxidoreductases	Catalyse oxidation or reduction reactions of substrates	Lipoxygenase, polyphenol oxidases, peroxidases	★★
EC 2 Transferases	Catalyse the transfer of functional groups from one molecule to another	Transglutaminases	★
EC 3 Hydrolases	Catalyse the hydrolysis (i.e., breaking of bonds with the help of water) of substrates	Amylases, lipases, proteases	★★★
EC 4 Lyases	Catalyse the addition or removal of groups from substrates by non-hydrolytic mechanisms	–	×
EC 5 Isomerases	Intramolecular rearrangement of substrates	Glucose isomerase	★
EC 6 Ligases	Catalyse linking of substrates	–	×
EC 7 Translocases	Catalyse the movement of molecules across membranes	–	×

polygalacturonases). Exogenous enzymes have only desirable properties if they are carefully controlled. Examples of applications of such enzymes are lactase for lactose hydrolysis, rennin in cheese manufacturing, papain for tenderisation and meat recovery or amylases for improving bread volume.

Enzymes are classified based on their reaction mechanism into *seven main categories* (Table 3.4). It should be noted that not all classes are important in food technology that is mainly concerned with hydrolases and oxidoreductases. At the time of writing this sentence, more than 6500 enzymes are known to science and make their classification a challenging task. The *enzyme commission numbering system* (EC number) classifies enzymes based on the chemical reactions they catalyse using a unique four-digit numerical code separating each digit by a full stop, which provides a finer distinction of their catalytic activity. For example, in the code EC 3.2.1.1, number 3 is the *main class* of the enzyme, a hydrolase. The second number is the *subclass*, a glycosylase that hydrolyses glycosidic bonds. The third number is the *sub-subclass*, a glycosylase that hydrolyses *O*- and *S*-glycosyl compounds. Finally, the fourth number is the *serial number* of the enzyme within its sub-subclass and identifies the name of the individual enzyme, which in this case is α-amylase. The complete list of all enzymes can be found on online databases, such as the ExPASy Enzyme database.

Certain enzymes that hydrolyse proteins or polysaccharides may be *endo-* or exoacting enzymes (Fig. 3.27). Endoacting enzymes hydrolyse in the interior of the protein or polysaccharide and result in polypeptides or oligosaccharides. Exoacting

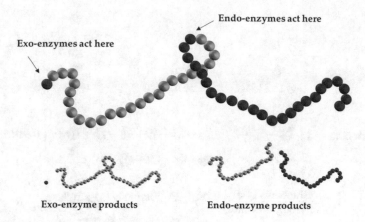

Fig. 3.27   Endo- and exoacting enzymes and their products

hydrolyse at the end of the chains and result in free amino acids or monosaccharides. A combination of both may lead to complete hydrolysis of the macromolecule, such as in glucose syrup production from starch, where a cocktail of enzymes is used to hydrolyse starch completely.

3.4.4   Food Enzymes

Most of the enzymes used for industrial purposes are of microbial origin produced through controlled fermentation and isolation of the desired enzyme. Bakery, dairy, and beverage industries are the most prominent enzyme users. Nevertheless, all sectors of the food industry are concerned with enzymes either directly or indirectly. The most important enzymes used in food processing and their specific activity is discussed below.

3.4.4.1   Carbohydrate Acting Enzymes

α-Amylases are endoenzymes that hydrolyse randomly α-$(1 \rightarrow 4)$ linkages to liquefy starch, but they do not hydrolyse α-$(1 \rightarrow 6)$ linkages (branching points). Products of α-amylase are maltose, glucose and dextrins. *The difference between dextrins and maltodextrins* is that dextrins are low-molecular-weight carbohydrates with α-$(1 \rightarrow 4)$ and α-$(1 \rightarrow 6)$ bonds, whereas maltodextrins have predominantly α-$(1 \rightarrow 4)$. As a result of branching, dextrins have quite different functional properties than maltodextrins. *β-Amylases* are exoenzymes that remove maltose units from non-reducing ends of amylose chains. In amylopectin, the cleavage stops a few units

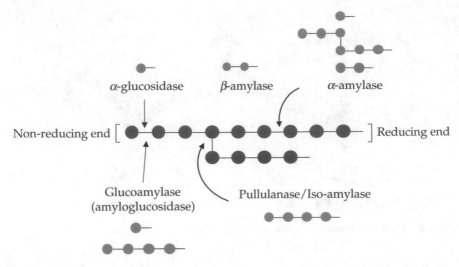

Fig. 3.28 Location of action and products of the four major amylases. Refer to the text for a description of their action

away from the α-(1 → 6)-branches. Glucoamylase (or amyloglucosidase) removes successive glucose units from the non-reducing ends of starch molecules by cleavage of the α-(1 → 4)-linkages and α-(1 → 6)-linkages. Unlike α-amylase, it hydrolyses linear and branched chains to form glucose. *Pullulanases* and iso-amylases hydrolyse only α-(1 → 6) glycosidic bonds in starch whereas α-glucosidases hydrolyse α-(1 → 4)-linkages from the non-reducing end to release glucose (Fig. 3.28). A typical application of amylases is in glucose, maltose, or high fructose corn syrups manufacturing. Hydrolysis starts with a corn starch suspension heated above the starch gelatinisation temperature (>70 °C). The pH of gelatinised starch is then adjusted to 6.5, and the paste is heated to 85 °C with the addition of α-amylase producing dextrins. Depending on the product, glucoamylase is then added to make glucose syrup or a mixture of pullulanases and β-amylases to make maltose syrup. Finally, if *high fructose corn syrup* (HFCS) is needed, glucose isomerase is used to convert glucose to fructose. Amylases are used in the baking industry to mildly hydrolyse starch to glucose used by yeasts to produce CO_2 essential in bread proofing. They are also added to bakery products to minimise staling since they hydrolyse amylose, thus delaying retrogradation. Amylases also reduce dough viscosity, improve loaf volume, crumb softness, and crust colour. In brewing, naturally occurring amylases in barley malt hydrolyse starch to maltose for yeast to utilise and produce CO_2 and ethanol.

Cellulases are enzymes that catalyse the hydrolysis of cellulose. *Endocellulases* randomly hydrolyse non-covalent bonds of crystalline cellulose to create free cellulose chains (i.e., not in crystalline form). *Exocellulases* cleave cellulose producing disaccharides (*cellobiose*) or tetrasaccharides (*cellotetraose*), and finally, *β-glucosidase (cellobiase)* hydrolyses cellobiose and cellotetraose into glucose

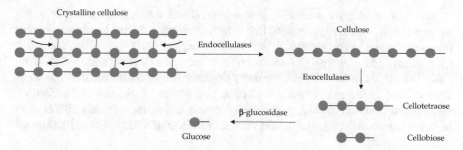

Fig. 3.29 Action of cellulases. Refer to the text for a description of their action

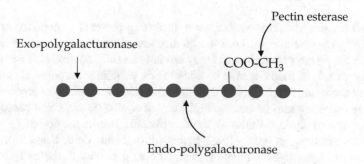

Fig. 3.30 Location of action and products of the pectinases. Refer to the text for a description of their action

(Fig. 3.29). Humans do not produce cellulases and are incapable of digesting cellulose (see Sect. 2.5.10, Dietary Fibre). It may, however, be partially broken down through fermentation in the large intestine producing *short-chain fatty acids* with beneficial properties to health.

Pectic enzymes break down pectin into oligosaccharides, galacturonic acid or de-esterify pectin. Pectin methylesterase hydrolyses the methoxyl group from pectin to form low-methoxyl pectin and polygalacturonic acid. Polygalacturonase (pectinase) hydrolyses α-(1 → 4) linkages of polygalacturonic acid to produce oligo-galacturonans and galacturonic acid. Polygalacturonase has both endo- and exo- forms that respectively hydrolyse pectin randomly or sequentially from the non-reducing ends (Fig. 3.30). Pectic enzymes are used in several food processing operations, such as in fruit juice and wine clarification or tomato processing. For instance, tomato product processing proceeds either by crushing tomatoes under high (~95 °C, *hot-break*) or low temperatures (~65 °C, *cold-break*). The main difference between these two processes is that the cold-break technology achieves a partial, whereas hot-break a total deactivation of pectic enzymes. Hot-break results in a paste with high viscosity, as pectin is intact and enhances thickness, suitable for sauces, ketchup, or purées. Cold-break leads to low viscosity pastes suitable for tomato juices and sauces, as pectin has been hydrolysed.

Invertase hydrolyses sucrose to glucose and fructose to a product that is known as invert sugar. Invert sugar is used in the confectionery industry because is sweeter than sucrose and has less tendency to crystallise (e.g., in soft candy fillings). Lactase (β-D-galactosidase) hydrolyses lactose to glucose and galactose and is used in the dairy industry to create *lactose-free* dairy products (milk, cream etc.). *Glucose oxidase* is used in the egg industry to deplete egg whites of glucose and reduce the potential of Maillard browning upon dehydration and storage. It may also be used in flours to induce the formation of disulfide links in gluten and improve flour bread-making capacity.

3.4.4.2 Protein Acting Enzymes

Proteases are enzymes responsible for hydrolysing proteins to peptides or amino acids. They are essential for both the quality and shelf life of foods, as they can influence the product at any stage if they are not carefully controlled. Proteases are naturally present in foods or may be added for a particular purpose. They may be classified according to their pH optima into *acid, neutral,* or *alkaline proteases,* and they may be both endo- or exo- acting. They may also be classified based on the nature of the catalytic residues at their active site into *aspartic-* (e.g., pepsins, cathepsins, rennins), *cysteine-* (e.g., papain, bromelain, ficin, actinidin), *serine-* (e.g., trypsin, chymotrypsin) *metallo-* (e.g., carboxypeptidase A, thermolysin), *threonine-, glutamic-,* and *asparagine-* proteases. When proteases hydrolyse from the *N*-terminus of the substrate-protein, they are called aminopeptidases. When they hydrolyse from the *C*-terminus, they are called carboxypeptidases.

Proteases may be added as *meat-tenderising* agents on inferior meat cuts (e.g., *papain*). The enzyme hydrolyses primarily connective tissue proteins (i.e., collagen) and softens the muscle. Other tenderising enzymes are *ficin* from figs, *bromelain* from pineapple and *microbial proteases.* Chymosin or rennin is found in the stomach of ruminants and is essential in the manufacturing of cheese. Chymosin acts on κ-casein and cleaves at the peptide bond between amino acid residues 105 (phenylalanine) and 106 (methionine). This results in *para κ casein,* which is hydrophobic and remains on the casein micelle, and *casein macropeptide,* which is hydrophilic and is transferred to the whey. In the presence of calcium, para-κ-casein and other molecules become now insoluble, and casein micelle remnants aggregate to form a gel, the *cheese curd,* which is the first step in many cheese manufacturing processes. *Digestive proteases* such as trypsin, chymotrypsin, and pepsin found in the animal pancreas may cause quality problems in muscle foods if contamination from intestines occurs (e.g., on minced products) that may lead to over-softening of meat. Proteases are also used to make protein hydrolysates with intense savoury taste used in snack products. They may be used to partially break down wheat flour proteins to reduce mixing time and make them more extensible. Some are added to help with flavour and texture development and speeding up fermentation in fermented dairy products. Finally, various strains of *proteolytic bacteria* that produce beneficial proteases may also be used during food fermentations to give foods unique flavour and textural characteristics.

3.4.4.3 Lipid Acting Enzymes

Lipases are hydrolases that hydrolyse triacylglycerols to form diglycerides, monoglycerides, glycerol, and free fatty acids (Fig. 3.31). Lipases have a dramatic impact on the quality of food products, especially those that contain significant amounts of fat. Lipases may lead to hydrolytic rancidity which is an undesirable reaction. For example, when free fatty acids are released in muscle foods, they react with proteins and denature them to give a tough texture. Furthermore, if they are not inactivated in milk, they release short-chain fatty acids that are very volatile, leading to the generation of off-flavours. However, lipases may have desirable effects when used in fermented products, as they are critical in cheese, salami, and other dry-meat product ripening. Short-chain fatty acids released from milk fat produce the characteristic odour and flavour of dairy products, especially those with eight carbon atoms. Finally, lipases can be used to modify the properties of lipids, especially in the margarine industry, to give different textures and melting points or to produce mono- and di-glycerides for use as emulsifiers.

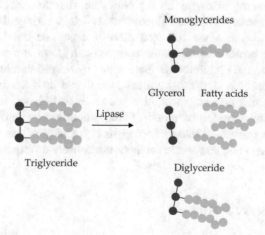

Fig. 3.31 Lipases hydrolyse triacylglycerols to form diglycerides, monoglycerides, glycerol, and free fatty acids

3.4.4.4 Other Enzymes

Lipoxygenases, peroxidases and *polyphenol oxidases* (PPO) are the most important representatives of oxidoreductases. *Lipoxygenases* are found in a wide variety of plants (primarily legumes) and have also been identified in animal tissues (e.g., in fish skin), and they catalyse the oxidation of unsaturated fatty acids. Linoleic acid (2 double bonds), linolenic acid (3 double bonds), and arachidonic acid (4 double bonds) are three naturally occurring lipoxygenase substrates. Lipoxygenases may have desirable outcomes when they are used for bleaching of wheat and soybean flours. They also contribute to the formation of disulfide bonds in the dough, thus avoiding the addition of chemical oxidisers. In other products, if the activity is not controlled, it may create significant quality defects, as lipid oxidation and reactivity of the breakdown products (hydroperoxides) give off-flavours and off-odours and may lead to textural problems. One of the purposes of blanching, an essential step before vegetable freezing, is to inactivate lipoxygenases and extend their shelf life.

Peroxidases oxidise several phenolic substrates (e.g., *p*-cresol, catechol, caffeic and coumaric acids), ascorbic acid, or aromatic amines. They are heat-stable enzymes responsible for the bleaching of chlorophylls and betalains, resulting in undesirable colour in canned vegetables. Reaction with *guaiacol* results in brown colour and the absence of brown colour is used as an indication of correct blanching. *Polyphenol oxidases* are discussed in detail separately in Chap. 5.

Naringinase is an *enzyme complex* with α-L-rhamnosidase and β-D-glucosidase activities. It is used as a debittering enzyme in citrus juice production, especially grapefruit, acting on naringin that gives grapefruit a bitter taste. α-L-Rhamnosidase acts first by breaking naringin into prunin and rhamnose, followed by β-D-glucosidase that breaks prunin into glucose and tasteless naringenin.

Transglutaminases catalyse the formation of a peptide bond between a free amine group of a protein-bound lysine and the acyl group at the end of the side chain of protein-bound glutamine. Bonds formed by transglutaminase exhibit high resistance to proteolysis. The most typical commercial products made with transglutaminases are imitation crabmeat, also known as surimi, and fish balls. A relatively unsuccessful commercial name of transglutaminase preparations creating confusion to both consumers and technologists is "meat glue", as the chemistry of adhesion (i.e., glues) and enzymatic catalysis are entirely different.

3.5   Learning Activities

3.5.1   Multiple-Choice Questions – Proteins

1) The isoelectric point of an amino acid is:

   (a) pH value at which the amino acid is neutral

(b) pH value at which the amino acid is positively charged

(c) pH value at which the amino acid is negatively charged

(d) pH value at which the amino acid is denatured

2) In L-amino acids, the amino group is positioned to the left of the chiral carbon. This statement is:

(a) False

(b) True

(c) It is true ONLY for non-polar and polar amino acids

(d) It is true ONLY for charged amino acids

3) A non-polar amino acid has

(a) A hydrophilic R-chain

(b) A hydrophobic R-chain

(c) A negatively charged R-chain

(d) A positively charged R-chain

4) A peptide bond is

(a) An amide bond between two amincs

(b) An amide bond between two amino acids

(c) A glycosidic bond between two amines

(d) A glycosidic bond between two amino acids

5) Primary structurc of a protein is:

(a) The sequence of monosaccharides in the polypeptide chain

(b) The sequence of fatty acids in the polypeptide chain

(c) The sequence of amino acids in the polypeptide chain

(d) None of the above

6) Secondary structure of a protein is:

(a) The specific geometrical arrangement of the polypeptide chain along one axis

(b) Two primary structures connected

(c) Two primary structures parallel to each other

(d) The specific arrangement of α-helix along a horizontal axis

7) Tertiary structure of proteins is stabilised by:

(a) Hydrophobic interactions, ionic bonds, hydrogen bonds and disulfide bridges

(b) Hydrophilic interactions, ionic bonds, hydrogenated bonds and trans bridges

(c) Hydrophilic interactions, ionic bonds, hydrogen bonds and disulfide bridges

(d) Hydrophobic interactions, ionic bonds, hydrogen bonds and glycosidic linkages

8) "Denaturation results in cleavage of peptide bonds":

(a) True for meat proteins ONLY

(b) False for meat proteins ONLY
(c) False for ALL proteins
(d) True for ALL proteins

9) Proteins usually unfold:

(a) Near their isoelectric point
(b) At extreme pH values, e.g., highly acidic or alkaline
(c) Near their isoelectric point but in the presence of NaCl
(d) At low to medium NaCl concentrations

10) "Protein hydrolysis involves breaking peptide bonds yielding smaller peptide chains or amino acids":

(a) True
(b) False
(c) True ONLY for fermented products
(d) True ONLY for acidic products such as orange juice

3.5.2   Multiple-Choice Questions – Enzymes

1) Enzymes, in general, act:

(a) on specific substrates catalysing a specific reaction
(b) on carbohydrates and proteins only catalysing a specific reaction
(c) on lipases only catalysing a range of hydrolytic reactions
(d) on plant and animal organisms only catalysing metabolic reactions

2) Complete the sentence: "A cofactor is a non-protein molecule…":

(a) that is bound to a polysaccharide and is required for hydrolysis
(b) that is bound to a protein and is required for the enzymatic action
(c) that is bound to the substrate and is required for the enzymatic action
(d) that is bound to a protein and is required for concurrent reactions

3) Enzymatic specificity is the ability of an enzyme:

(a) to catalyse two reactions at the same time
(b) to catalyse one reaction in proteins and one in polysaccharides
(c) to catalyse only one reaction
(d) to catalyse one reaction in each group of macronutrients (i.e., carbohydrates, proteins, lipids)

4) Enzymatic selectivity is:

(a) the ability of enzymes to select which reaction to carry out

(b) the preference of an enzyme to catalyse peptide instead of glycosidic bonds

(c) the industrial preference for an enzyme for a specific manufacturing process

(d) the preference of an enzyme toward similar competing substrates

5) Endo-acting enzymes:

(a) Act randomly in the interior of the protein or polysaccharide molecule

(b) Start hydrolysing from the outside of the protein or polysaccharide molecule

(c) Act in the endocarp of certain fruits leading to tissue softening

(d) Act in the exocarp of certain fruits leading to tissue firming

6) α-amylases catalyse:

(a) hydrolysis of ester bonds in triglycerides

(b) hydrolysis of α-1,4 linkages of starch

(c) hydrolysis of peptide bonds

(d) hydrolysis of α-1,6 linkages of starch

7) Pectin methyl-esterase hydrolyses the:

(a) acetyl group of pectin to form low-methoxyl pectin and polygalacturonic acid.

(b) methoxyl group of pectin to form high-methoxyl pectin and polygalacturonic acid

(c) methoxyl group of pectin to form low-methoxyl pectin and polygalacturonic acid

(d) methoxyl group of pectin to form low-methoxyl pectin and polyguluronic acid

8) Lactase catalyses hydrolysis of

(a) Lactose to glucose and fructose

(b) Lactic acid to glucose and galactic acid

(c) Galactose to glucose and lactose

(d) Lactose to glucose and galactose

9) Proteases catalyse

(a) Hydrolysis of peptide bonds of protein to peptides and amino acids

(b) Hydrolysis of peptide bonds of polysaccharides to monosaccharides

(c) Synthesis of peptide bonds to form proteins

(d) Synthesis of peptide bonds to form polysaccharides

10) Lipases hydrolyse:

(a) Fats to form diglycerides, monoglycerides and free fatty acids

(b) Lipose to glucose and liparose

(c) Fatty acids to triglycerides

(d) Triglycerides to cholesterol and fat-soluble vitamin D

3.5.3   Short Answer Questions – Further Reading

1) Draw and discuss the changes in the ionisation state of amino acids with changes in pH. What is the consequence of changes in pH for food proteins?

2) Draw cysteine and identify the carboxyl and amino groups, the side chain, and the sulfhydryl group. What kind of special bonds is this amino acid likely to form?

3) The amino acid sequence SRFRASHGDFRI is from the protein vicilin. Identify the amino acids in this sequence.

4) The table below shows the dissociation constants of three amino acids. Calculate their pI using the correct equation.

	pK_{COOH}	pK_{NH3}^+	pK_R
Serine	2.21	9.15	–
Aspartic acid	1.88	9.60	3.65
Histidine	1.82	9.17	6.00

5) The tripeptide LLF is a bitter peptide found in whey protein hydrolysates. Draw its structure.

6) The following protein structure is bovine α-lactalbumin:

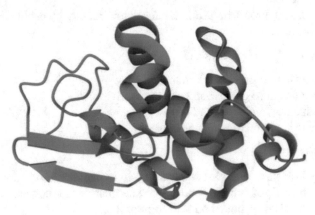

How many α-helixes and β-sheets can you count? Are β-sheets parallel or anti-parallel? Identify a loop and a random coil.

7) Name and discuss the forces that stabilise the tertiary structure of proteins.

8) List and discuss the factors that affect protein denaturation.

9) a) The following protein is hydrolysed with an enzyme that cleaves the bonds between R and F amino acids (shown in bold): RRHKNKNPFHFNSK**RF**QT LFKNQYGHVRVL. Write the sequence of the peptides that are formed.

b) The same protein is hydrolysed with the action of acid and heat at the amino acid positions shown in bold:

RRHKNKN**P**FHFNSK**R**FQTLFKN**QY**GHVR**VL**. Write the sequence of the peptides that are formed. Name the free amino acids that are released.

10) A protein has the following primary structure: LLVWVAVWEDRKH. You use this protein to make an oil-in-water salad dressing. Identify and explain which part of the protein arranges towards the oil phase and which part towards the aqueous phase (Tip: find the hydrophobic amino acids in the sequence).

11) A company uses protein gelation technology to stabilise the structure of a milk-based dessert. Describe the steps and the chemical changes that occur in proteins during gelation.

12) Discuss the differences between the terms "gluten", "glutenin", "gliadin", "glutelin" and "prolamin".

13) Name and discuss the factors that control enzymatic activity.

14) A company wants to make tomato paste with low viscosity suitable for soup formulations. **Immediately** after tomato crushing, they pasteurise the pulp to 88 °C and unfortunately, the paste has high viscosity making it unsuitable for the end-product. Explain what happened, which enzymes act on tomato pulp and how you will resolve this problem.

15) Identify the locations of action and products of α-amylase, β-amylase, and glucoamylase. Describe the use of these enzymes in glucose syrup production.

16) Describe the action of chymosin, invertase, and lactase on their respective substrates. Give one food application of each enzyme.

17) **Online activity** Visit the website https://www.expasy.org/resources/enzyme and find the enzymes EC 3.4.22.2, EC 3.1.1.3, and EC1.13.11.12. Browse the information shown.

18) **Further reading**: Find the review article "McClements, D. J. (2004). Protein-stabilized emulsions. Current Opinion in Colloid & Interface Science, 9(5), 305-313".

   • Name physical and chemical parameters that influence emulsion stability.
   • How do thermal processing, homogenisation, drying, and freezing influence emulsion stability?
   • What is the influence of pH and ionic strength on emulsion stability?

19) **Further reading**: Find the book "Whitaker, J. R., Voragen, A. G. J., & Wong, D. W. S. (2003). Handbook of food enzymology. New York: Marcel Dekker."

- Find the chapters of the following enzymes and identify their substrates and products

Horseradish peroxidase
Glucose oxidase
Alcohol dehydrogenase
Amylosucrase
Chlorophyllase
Phytase
Lysozyme

- Find the chapter "Proteolytic Enzymes" and identify:

The uses of proteolytic enzymes in the food industry
The types of proteolytic enzymes
Name three methods of measuring proteolytic activity

3.5.4 Fill the Gaps

1) _____ amino acids are those that the body does not synthesise and must be obtained from the diet.

2) Isoelectric point is the ____ at which amino acids have _____ net charge.

3) The sulfhydryl groups of two _____ oxidise easily to form _____.

4) The bond between two sulfhydryl groups is called _____ _____.

5) Schiff base is a compound that contains a _____ double bond.

6) _____ is used in the detection and quantification of free amino acids.

7) _____ and _____ are the major proteins found in beans, peas, and other legumes.

8) Secondary structure results in _____ of the polypeptide chain leading to a unique _____.

9) There are two common types of tertiary structures: _____ and _____.

10) Denaturation is caused by _____ or _____ *agents*.

11) Detergents are denaturing agents that disrupt _____ _____.

12) Another term for the "Hofmeister series of salts" is "_____ series of salts".

13) In emulsification, proteins create a protective _____ around the _____ droplets.

14) Enzymes decrease the _____ _____ that is required thus facilitating the reaction.

15) The ability of an enzyme to catalyse only one reaction is termed _____.

16) Enzymes with similar enzymatic activities but different protein structures and are called _____.

17) Enzymes that require a metal in their structure for activity are known as _____.

18) _____ Michaelis-Menten constant values correspond to a high affinity for the substrate.

19) In the enzyme code EC 3.2.1.1, number 3 is the _____ _____ of the enzyme.

20) _____ enzymes hydrolyse in the interior of the protein or polysaccharide.

21) When proteases hydrolyse from the N-terminus of the substrate-protein they are called _____ whereas when they hydrolyse from the C-terminus they are called _____.

22) Peroxidases oxidise _____ substrates.

23) Surimi products are made with the action of _____.

24) Naringinase is an enzyme complex that is used as a _____ enzyme.

Chapter 4
Lipids

Learning Objectives

After studying this chapter, you will be able to:

- Describe the structure of food lipids
- Describe the mechanism of lipid oxidation
- Discuss the factors that influence lipid oxidation
- Describe fat crystal formation
- Describe triacylglycerol conformation and crystal polymorphism
- Discuss factors that influence the melting point of fat crystals

4.1 Introduction

Lipids are substances soluble in organic solvents but insoluble in water. They are major components of adipose tissue and the cell membranes of animals. Their unique physical and chemical properties make them essential contributors to the sensory properties, storage stability and overall food quality, especially those rich in lipids. Lipids are composed of eight groups with distinct physical, chemical, and physiological properties. These groups are the *fatty acids*, *triacylglycerols*, *phospholipids*, *sphingolipids*, *steroids*, *waxes*, *fat-soluble vitamins*, and *carotenoids*. *Fats* and *oils* in foods are differentiated depending on their source and physical state at room temperature. *Fats* are solid at room temperature (~ 20 °C), while *oils* are liquid. Fats are usually derived from animals and are primarily saturated. In contrast, oils are derived from plants and are mono- or polyunsaturated, as we will see in the following sections. Three exceptions are coconut and palm oils, derived from plants but are solid at room temperature, and fish oils, which at room temperature are liquid. Animal sources include meat, poultry, and dairy, whereas plant sources include nuts, seeds, avocado, olives, and coconut. Most fruits and vegetables contain very little fat. In the following sections, we explore the structure and properties of these molecules and their relevance to food manufacturing, quality, and shelf life.

4.2 Fatty Acid Nomenclature and General Characteristics

Fatty acids are carboxylic acids with an *aliphatic* (straight) carbon chain. Fatty acids differ from one another in two major ways: *the length of the carbon chain* and the *degree of saturation*. The number of carbon atoms determines the length of the chain, and in naturally occurring fatty acids, it is always an *even number*, usually between 4 and 22 carbon atoms. The one end of the structure is hydrophilic, as it contains the carboxylic group, whereas the hydrocarbon chain is hydrophobic. The counting of carbon atoms starts from the carbon atom of the carboxyl group and proceeds to the left (Fig. 4.1). The *degree of saturation* is determined by the number of double bonds between carbon atoms. In the absence of a carbon-carbon double bond, the fatty acids are called saturated. Fatty acids with one carbon-carbon double bond are called monounsaturated, and with two or more carbon-carbon double bonds are called polyunsaturated. Saturated fatty acids are primarily found in animal sources such as meat, poultry, butter, egg yolk, or lard. Plant sources that contain saturated fatty acids are coconut and palm oils. Major sources of monounsaturated fatty acids are olive and avocado oils and peanut butter. Polyunsaturated fatty acids are predominantly found in plants (e.g., corn, canola, sunflower, or flaxseed) and fish oils.

Fatty acids can be named according to the systematic IUPAC nomenclature system or use empirical names that usually relate to its source. For example, the

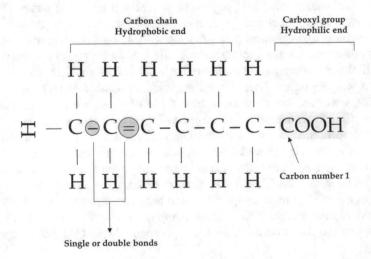

Fig. 4.1 Structure of fatty acids. See text for description

Table 4.1 Saturated fatty acids found in foods

Common name	Systematic name	Abbreviation	Sources
Butyric acid	Butanoic acid	C4:0	Butter
Caproic acid	Hexanoic acid	C6:0	Coconut, palm oil, butter
Caprylic acid	Octanoic acid	C8:0	Coconut, palm oil, butter
Capric acid	Decanoic acid	C10:0	Coconut, palm, nut oil, butter
Lauric acid	Dodecanoic acid	C12:0	Coconut, palm, nut oil, butter
Myristic acid	Tetradecanoic acid	C14:0	Animal and plant fats
Palmitic acid	Hexadecanoic acid	C16:0	Animal and plant fats
Stearic acid	Octadecanoic acid	C18:0	Animal fats
Arachidic acid	Eicosanoic acid	C20:0	Peanut oil

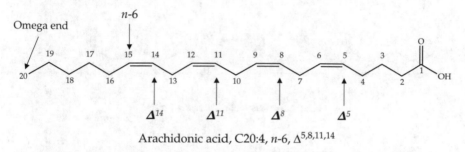

Arachidonic acid, C20:4, n-6, $\Delta^{5,8,11,14}$

Fig. 4.2 Example of notations in fatty acids using the unsaturated arachidonic acid. Its IUPAC name is 5,8,11,14-eicosatetraenoic acid

empirical name of butanoic acid is butyric acid, as it is found in butter. A way of writing fatty acids is by using numerical abbreviations that show the number of carbon atoms and the double bonds (Table 4.1). For example, butyric acid is 4:0 because it has four carbon atoms and zero double bonds, whereas 18:1 represents oleic acid with 18 carbon atoms and one double bond.

Omega fatty acids (ω) are polyunsaturated fatty acids characterised by the presence of a double bond three (ω-3) or six (ω-6) atoms away from the terminal methyl group (Fig. 4.2 and Table 4.2). Examples of ω-3 are α-linolenic acid (C18:3 n-3) or eicosapentaenoic acid (C20:5 n-3) and examples of ω-6 are linoleic acid (C18:2 n-6) or arachidonic acid (C20:4, n-6). The letter n in the abbreviation indicates the carbon atom number where the first double bond is found, counting from the end of the structure. The notation $\Delta^{x,y,\cdots}$ is also used to indicate the positions of double bonds. For instance, arachidonic acid is $\Delta^{5,8,11,14}$, indicating double bonds on carbons 5, 8, 11, and 14. In Δ notation, counting starts from the carboxyl group (Fig. 4.2).

The presence of double bonds in fatty acids results in cis- or trans- isomerism. In the *cis*- configuration, hydrogens attached to carbon atoms are on the same side of the double bond, whereas in *trans*- configuration, they are on either side of the double bond (Fig. 4.3). Fatty acids in *trans*- configuration have been found to have

Table 4.2 Unsaturated fatty acids found in foods

ω-n	Common name	Abbreviation	Δ^n	Configuration	Common sources
ω-3	α-Linolenic acid	C18:3	$\Delta^{9,12,15}$	cis	Flaxseeds, chia, walnuts
ω-3	Eicosapentaenoic acid	C20:5	$\Delta^{5,8,11,14,17}$	cis	Fish
ω-6	Linoleic acid	C18:2	$\Delta^{9,12}$	cis	Peanut oil, olive oil
ω-6	Linolelaidic acid	C18:2	$\Delta^{9,12}$	trans	Partially hydrogenated oils
ω-6	Arachidonic acid	C20:4	$\Delta^{5,8,11,14}$	cis	Meat, eggs
ω-7	Palmitoleic acid	C16:1	Δ^9	cis	Macadamia nuts
ω-9	Oleic acid	C18:1	Δ^9	cis	Olive oil, canola oil
ω-9	Elaidic acid	C18:1	Δ^9	trans	Hydrogenated oils
ω-9	Erucic acid	C22:1	Δ^{13}	cis	Mustard oil

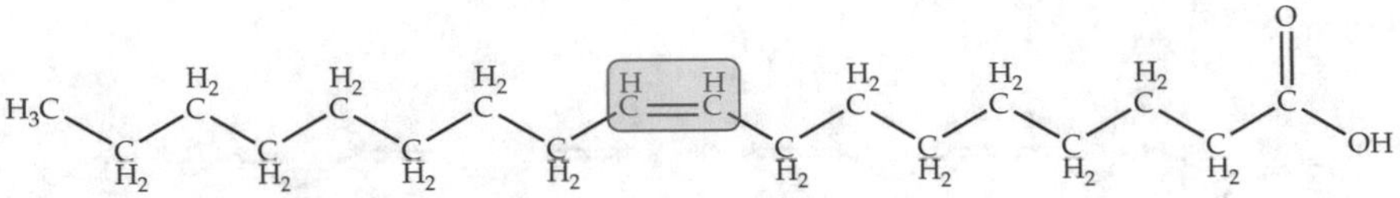

Oleic acid (*cis*-octadecenoic acid)

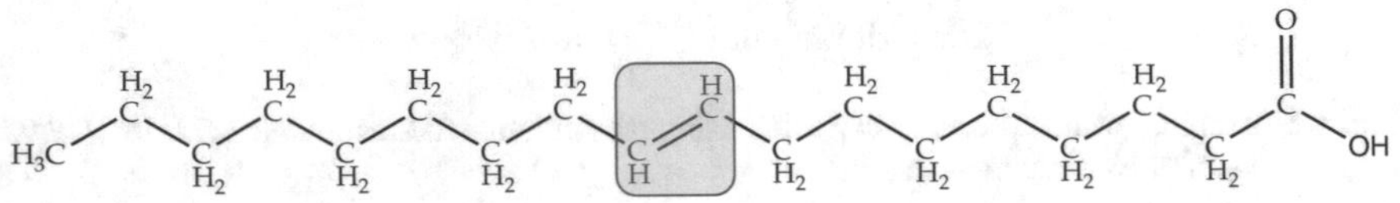

Elaidic acid (*trans*-octadecenoic acid)

Fig. 4.3 *Cis-, trans-* isomerism of fatty acids. In the blue box, the positioning of hydrogen atoms around the double bond is emphasised

$$-C-C=C-C=C-C=C-C-$$

Fig. 4.4 In conjugated double bonds, single bonds separate double bonds

negative implications to health and are usually present because of certain processing steps (e.g., hydrogenation or frying). However, modern processing technologies have alleviated this problem significantly.

A structural characteristic of some fatty acids is the conjugated double bond when at least one pair of double bonds is separated by a single bond (Fig. 4.4). The presence of conjugated double bonds is important, as it is responsible for colour in carotenoids and the properties of the family of *conjugated linoleic acids* (more than 28 isomers) with beneficial health effects (e.g., weight management, combating cancer etc.). It should be noted that most naturally occurring fatty acids are in *cis*-configuration, *not* conjugated, and with *even-numbered* carbon atoms.

4.3 Triacylglycerols (TAGs)

Esterification is the reaction of an alcohol with a carboxylic acid to give ester and water. Fats and oils are chemically known as *triacylglycerols* (*TAGs*) and are formed by the *reaction of a fatty acid with glycerol* (Fig. 4.5). When three fatty acids are attached to glycerol, they form a *triacylglycerol*; two fatty acids form a diacylglycerol, whereas one fatty acid forms a monoacylglycerol. The fatty acids on the glycerol can be identical (*simple triacylglycerol*) or different (*mixed triacylglycerol*) (Fig. 4.6). When the middle carbon in glycerol has four different substituent groups, it becomes chiral with mono-, di-, and mixed TAGs having a chiral carbon. To distinguish the configuration of the fatty acids in TAGs, the carbon atoms of glycerol are numbered *stereospecifically*. A Fischer projection must be drawn with the hydroxyl group at C-2 to the left. The carbon atom at the top of the structure is C-1 and at the bottom is C-3. In this case, the prefix *sn* (*stereospecific numbering*) is used and in that way, it is possible to differentiate the positioning of the fatty acid(s) (Fig. 4.7). For instance, the names 1-oleoyl-2-stearoyl-3-palmitoyl-*sn*-glycerol, *sn*-1-oleo-2-stearo-3-palmitin, and *sn*-glycerol-1-oleate-2-stearate-3-palmitate are all equivalent referring to the same TAG and allow to distinguish the position of the esterified fatty acid(s). In this example, the -OH group of glycerol at C-1 is esterified with oleic acid, at C-2 with stearic acid, and at C-3 with palmitic acid. If we just write "oleoyl (O) – stearoyl (S) – palmitoyl (P) – glycerol", the positioning of the esterified fatty acids is ambiguous, as there are six possible permutations (i.e., OSP, OPS, SOP, SPO, POS, and PSO). It is important to know the stereospecific numbering in reactions such as interesterification or when crystallisation of TAGs is considered, as the location of the fatty acid on glycerol plays an important role in the physical properties of the resulting TAGs. The fatty acid composition of foods is quite variable and complex, but some general guidelines may be drawn. Vegetable oils from oilseeds are highly unsaturated with 18-carbon fatty

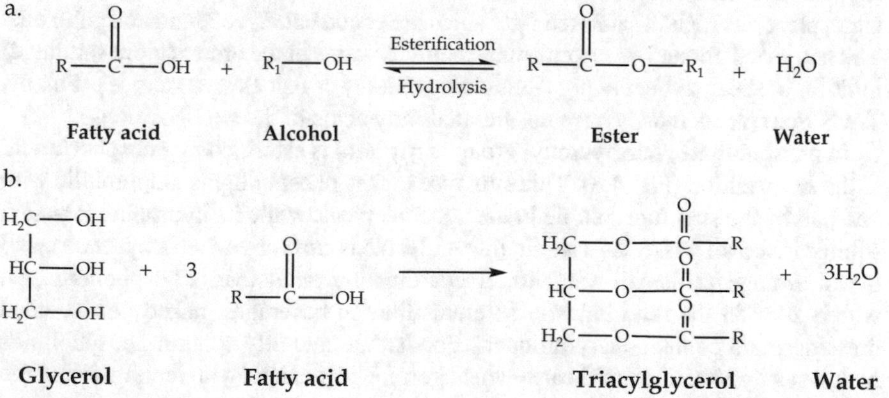

Fig. 4.5 (a) General description of esterification, and (b) esterification of glycerol with a fatty acid to form triacylglycerol

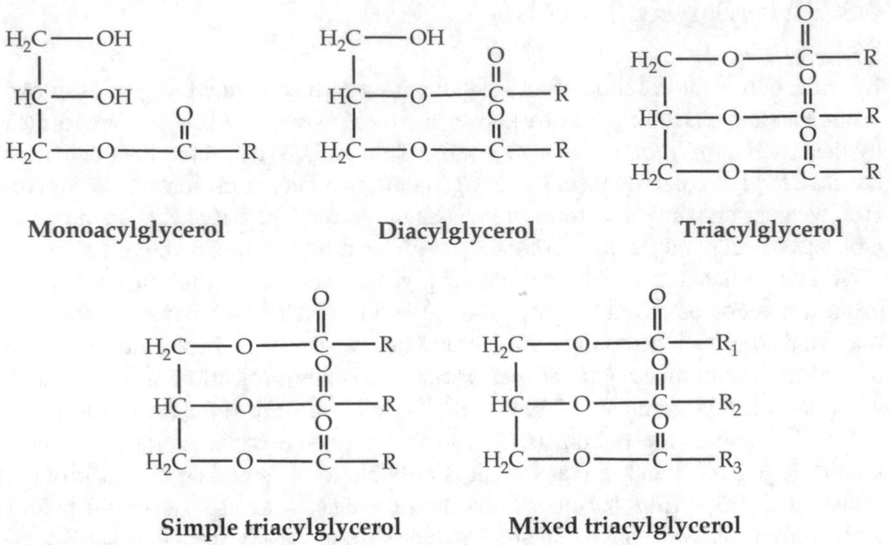

Figures (Fig. 4.6):

$$H_2C\!-\!OH$$
$$|$$
$$HC\!-\!OH$$
$$|$$
$$H_2C\!-\!O\!-\!\overset{\overset{O}{\|}}{C}\!-\!R$$

Monoacylglycerol

$$H_2C\!-\!OH$$
$$|$$
$$HC\!-\!O\!-\!\overset{\overset{O}{\|}}{C}\!-\!R$$
$$|$$
$$H_2C\!-\!O\!-\!\overset{\overset{O}{\|}}{C}\!-\!R$$

Diacylglycerol

$$H_2C\!-\!O\!-\!\overset{\overset{O}{\|}}{C}\!-\!R$$
$$|$$
$$HC\!-\!O\!-\!\overset{\overset{O}{\|}}{C}\!-\!R$$
$$|$$
$$H_2C\!-\!O\!-\!\overset{\overset{O}{\|}}{C}\!-\!R$$

Triacylglycerol

$$H_2C\!-\!O\!-\!\overset{\overset{O}{\|}}{C}\!-\!R$$
$$|$$
$$HC\!-\!O\!-\!\overset{\overset{O}{\|}}{C}\!-\!R$$
$$|$$
$$H_2C\!-\!O\!-\!\overset{\overset{O}{\|}}{C}\!-\!R$$

Simple triacylglycerol

$$H_2C\!-\!O\!-\!\overset{\overset{O}{\|}}{C}\!-\!R_1$$
$$|$$
$$HC\!-\!O\!-\!\overset{\overset{O}{\|}}{C}\!-\!R_2$$
$$|$$
$$H_2C\!-\!O\!-\!\overset{\overset{O}{\|}}{C}\!-\!R_3$$

Mixed triacylglycerol

Fig. 4.6 Different types of glycerides formed from the esterification of glycerol with fatty acids

Fig. 4.7 Fischer projection of a TAG. Stereospecific numbering of triglycerides is used to designate the configuration of triacylglycerides

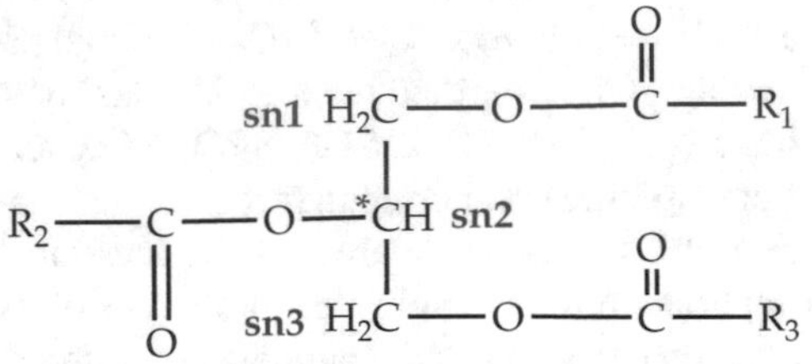

$$sn1\ H_2C\!-\!O\!-\!\overset{\overset{O}{\|}}{C}\!-\!R_1$$
$$|$$
$$R_2\!-\!\overset{\|}{\underset{O}{C}}\!-\!O\!-\!\overset{*}{C}H\ sn2$$
$$|$$
$$sn3\ H_2C\!-\!O\!-\!\overset{\overset{O}{\|}}{C}\!-\!R_3$$

acids. Oils high in oleic acid include olive and canola oils, high in linoleic acid are soybean and corn oils, whereas high in linolenic acid is linseed oil. On the other hand, plant oils high in saturated fatty acids are cocoa butter, coconut, and palm oils. The levels of saturated fats from animal sources varies in the order: (more saturated) milk fat > sheep > beef > pig > chicken > turkey > fish (less saturated). Finally, TAGs from plants mostly have unsaturated fatty acids in the sn-2 position.

In phospholipids, one hydroxyl group in glycerol is esterified by phosphoric acid at the sn3 position (Fig. 4.8). This structure makes phospholipids amphiphilic with one part of the structure soluble in the aqueous phase while its hydrophobic part is soluble in the oil phase. As a result, this molecule is *surface-active* (*surfactant*), and it can arrange at the oil-water interfaces creating emulsions. Phospholipids are widely used in the food industry as emulsifiers in beverages, baked goods, salad dressings, and confectionery products. Foods that naturally contain phospholipids include egg yolks, liver, soybeans, wheat germ, and peanuts, with lecithin being the most characteristic example.

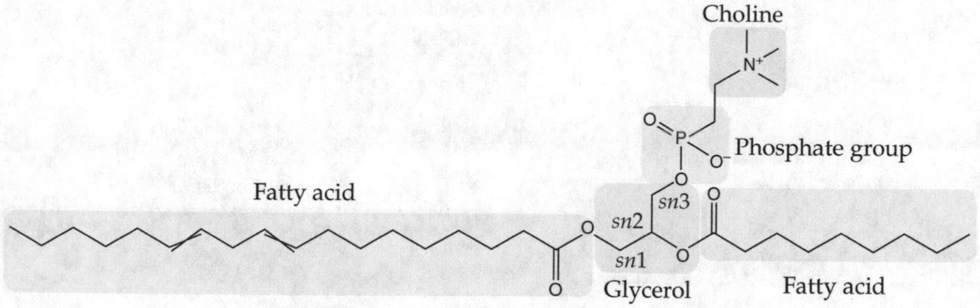

Fig. 4.8 Structure of lecithin, a common food phospholipid

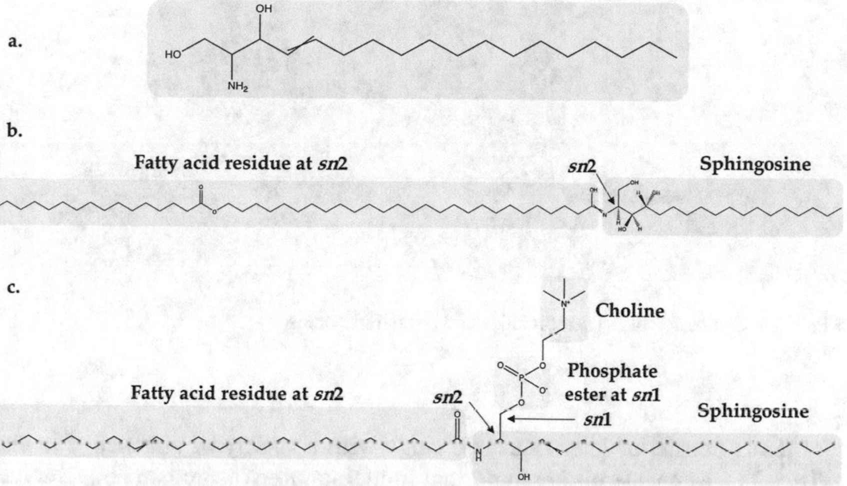

Fig. 4.9 Structures of (**a**) sphingosine, (**b**) ceramide, and (**c**) sphingomyelin

Sphingolipids are a class of lipids containing a backbone of *sphingoid bases*. Sphingosine is an amino alcohol that forms the structural basis of the most important sphingolipids (Fig. 4.9). For example, ceramides are composed of sphingosine with an *N*-linked fatty acid at the *sn*-2 position. Derivatives of ceramides may be formed by reactions of the -OH group at *sn*-1. For example, sphingomyelin consists of a ceramide with *phosphocholine* esterified at *sn*-1. When a single sugar residue, usually glucose or galactose, is attached at *sn*-1, a cerebroside is obtained, whereas when an oligosaccharide with one or more *sialic acids* is attached a ganglioside is obtained (Fig. 4.10). These compounds play an important role in signal transduction and cell regulation and although they are not typically considered as dietary lipids due to their low concentration their role is being continuously examined and re-evaluated. Foods rich in sphingolipids and their derivatives include dairy, meat, eggs, soy, and seafood (e.g., mussels, scallops etc.), whereas fruits and vegetables generally have a low content of these compounds.

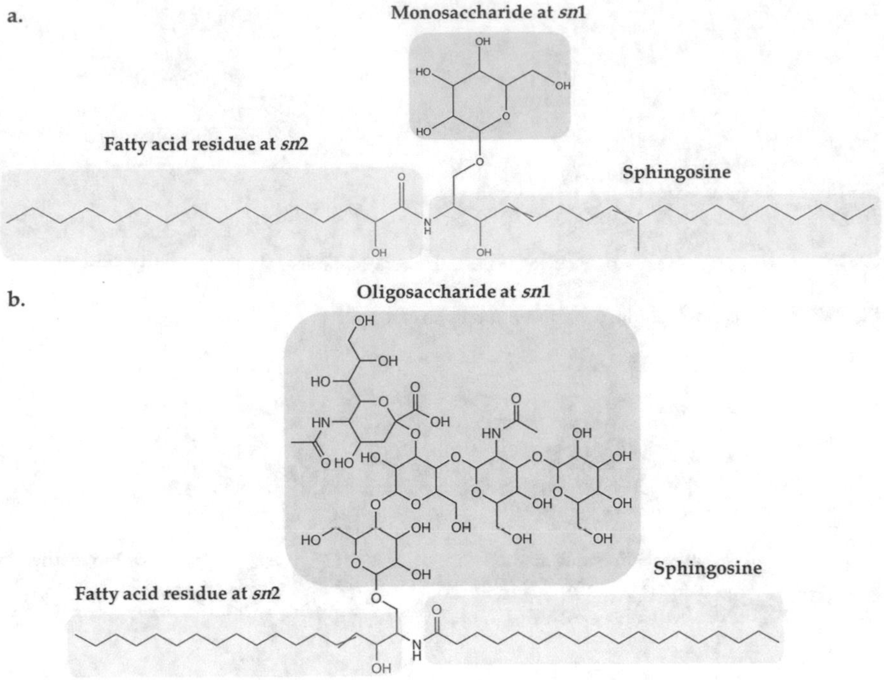

a.

Monosaccharide at *sn*1

Fatty acid residue at *sn*2

Sphingosine

b.

Oligosaccharide at *sn*1

Sphingosine

Fatty acid residue at *sn*2

Fig. 4.10 Structures of (**a**) cerebrosides and (**b**) gangliosides

Sterols consist of interconnected rings with a variety of attached side chains (Fig. 4.11). Many sterols are important in maintaining the human body, including cholesterol, bile, sex hormones (i.e., testosterone, oestrogen), adrenal hormones (i.e., cortisol), and vitamin D. The dietary sterol of the greatest significance in foods is cholesterol, where elevated levels in the blood are associated with diseases such as atherosclerosis that increases the risk of heart attack or stroke. Major dietary sources of cholesterol include red meat, egg yolks, liver, and butter. Phytosterols or stanolsare plant sterols (e.g., *β*-sitosterol or stigmasterol) that may compete with intestinal cholesterol absorption and reduce its levels in the blood. Products belonging to the broader group of *functional foods* are available in the market, incorporating phytosterols in the formulations (e.g., margarine or yoghurt drinks) to help with cholesterol reduction. Finally, *waxes* are mixtures of esters of fatty acids and long-chain alcohols but are not triacylglycerides. *Carnauba wax*, *beeswax*, and *candelilla wax* are typical examples of glazing agents and fruit waxing in the food industry.

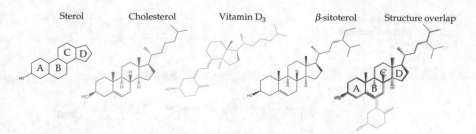

| Sterol | Cholesterol | Vitamin D₃ | β-sitoterol | Structure overlap |

Fig. 4.11 Structures of sterols. The last image shows a structure overlap to illustrate common structural aspects between them. The common characteristic is the *gonane* structure consisting of three molecules of cyclohexane (A, B, and C rings) and one of cyclopentane (D ring)

4.4 Lipid Oxidation

Lipid oxidation, also known as *oxidative rancidity*, is the reaction of lipids with oxygen in a generally detrimental process for foods. Food eventually oxidises and the objective of the food scientist is to *delay oxidation until the intended shelf life of the product*. Three types of lipid oxidation may be distinguished: auto-oxidation (triplet O_2 oxidation), photo-oxidation (singlet O_2 oxidation) and *enzymatic oxidation* (lipoxygenase). Autoxidation of lipids proceeds by a *free radical mechanism*. Free radicals are molecules or atoms with unpaired electrons and can oxidise easily other compounds. Auto-oxidation takes place in three stages: *initiation, propagation,* and *termination* (Fig. 4.12).

Initiation of oxidation occurs when a hydrogen atom in the double bonds of unsaturated fatty acids is removed to form an alkyl radical (R˙). Metal ions (usually Fe, Cu, Mg, or Ni), UV radiation (light), singlet oxygen, lipoxygenase, and high temperatures may cause or promote oxidation. *Pro-oxidants* accelerate oxidation by direct interaction with unsaturated fatty acids to form lipid hydroperoxides or by promoting the formation of free radicals. *Propagation* involves the addition of oxygen to the alkyl radical and the production of *peroxyl radicals* (ROO˙). Peroxyl radicals react with another fatty acid to produce *hydroperoxides* (ROOH) and a new alkyl radical, beginning a new oxidation cycle. *Termination* occurs when free radicals interact with each other or with an antioxidant and result in unreactive molecules. Hydroperoxides decompose to a multitude of compounds with flavour (aldehydes, ketones, hydrocarbons, furans, acids) and biological significance causing major quality deterioration, including generation of off-flavours in oils, changes in melting point in TAGs, formation of coloured products, formation of toxic compounds or loss of nutritional value.

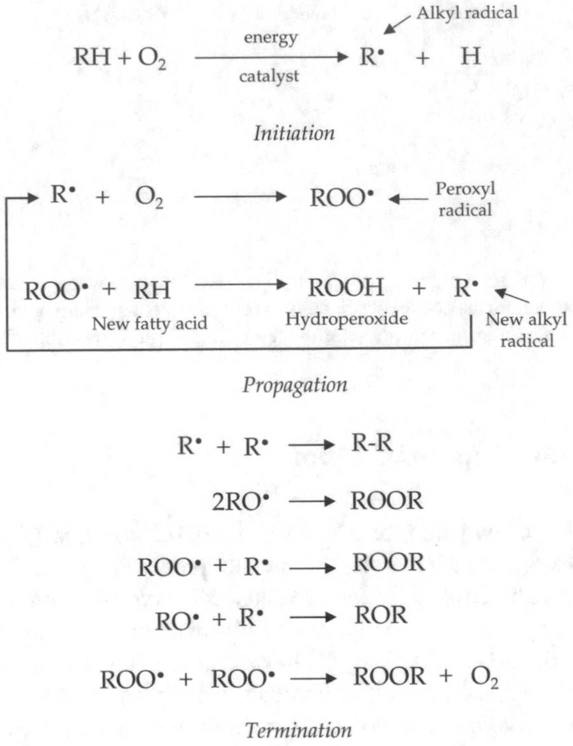

Fig. 4.12 The three stages of lipid oxidation. See text for description

$$RH + O_2 \xrightarrow[\text{catalyst}]{\text{energy}} R^\bullet + H$$

Alkyl radical

Initiation

$$R^\bullet + O_2 \longrightarrow ROO^\bullet$$

Peroxyl radical

$$ROO^\bullet + RH \longrightarrow ROOH + R^\bullet$$

New fatty acid   Hydroperoxide   New alkyl radical

Propagation

$$R^\bullet + R^\bullet \longrightarrow R\text{-}R$$

$$2RO^\bullet \longrightarrow ROOR$$

$$ROO^\bullet + R^\bullet \longrightarrow ROOR$$

$$RO^\bullet + R^\bullet \longrightarrow ROR$$

$$ROO^\bullet + ROO^\bullet \longrightarrow ROOR + O_2$$

Termination

Singlet Oxygen Oxidation (Photo-Oxidation)
There are two types of O_2, triplet ($^3O_2$) and singlet ($^1O_2$) oxygen. The difference lies in the arrangement of electrons in the molecular orbitals of O_2. The full explanation is quite involved, and details on the different types of O_2 can be found in a general chemistry textbook. What is relevant to our discussion is that $^1O_2$ is more reactive than $^3O_2$. Singlet O_2 oxidation requires light and the presence of photosensitisers. Photosensitisers are compounds that absorb light and transfer the energy to $^3O_2$ converting it to $^1O_2$. Singlet oxygen then reacts directly with the double bonds of fatty acids, and hydroperoxides are formed. Photosensitisers (e.g., chlorophyll and riboflavin) are naturally found in raw ingredients, and in most cases, it is not possible to remove them from food.

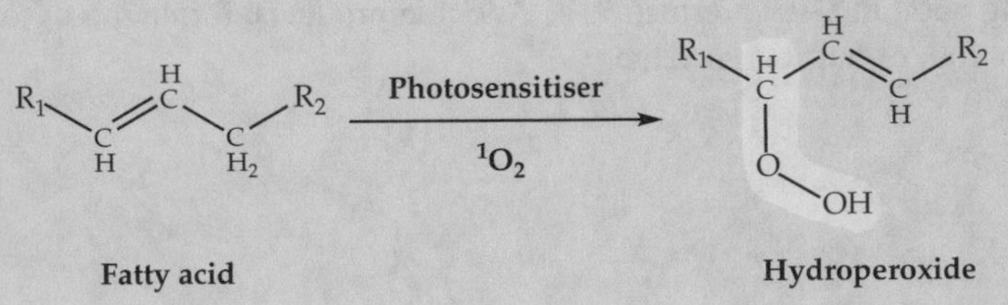

Fatty acid   **Hydroperoxide**

The main differences between autooxidation and photo-oxidation are that $^1O_2$ oxidation **does not proceed through radical formation and does not have an induction period**. Additionally, photo-oxidation is not particularly influenced by temperature, the degree of unsaturation of fatty acids, and the amount of oxygen dissolved in food. Photo-oxidation produces different hydroperoxides than autooxidation that decompose to distinct breakdown products (e.g., aldehydes). Analysis of breakdown products allows differentiation between the two mechanisms and identifies which type of oxygen is responsible for product oxidation. **Antioxidants are not very effective against $^1O_2$ oxidation**. Packaging impermeable to light, vacuum packaging, removal of photosensitisers or addition of singlet oxygen quenchers (e.g., ascorbic acid, tocopherol, or carotenoids) are some strategies that can be employed to minimise photo-oxidation depending on product composition. Soybean oil and milk are two foods that need to be stored away from sunlight because they are prone to photo-oxidation.

Water activity is an important factor influencing the rate of lipid oxidation. The relationship between a_w and oxidation rate is quite complex, and each food should be regarded separately. Generally, the oxidation rate is: fast at very low a_w (< 0.2), low at intermediate a_w (0.2–0.5), and fast at high a_w (> 0.5) (see also Fig. 1.12). Catalyst activity (e.g., Fe, Cu, Mg, or Ni) is high in the dry state of food, supporting fast reactions. As water increases, it hydrates catalysts, thus reducing their catalytic action, and the rate of lipid oxidation slows down. In addition, at intermediate a_w water interacts with polar hydroperoxides (ROOH) produced in the propagation step and cannot react fast enough. The increase of oxidation rate at high a_w (> 0.5) (Fig. 1.12) is usually due to the increased re-mobilisation of the catalysts and the reduced viscosity of the aqueous phase that both facilitate the reaction. The increased rate is *not* due to the increased oxygen concentration in the aqueous phase since the reaction occurs in the lipid phase. It should also be noted that the solubility of oxygen in lipids is multiple times greater than in water. *In foods susceptible to lipid oxidation, control of water activity and temperature is necessary.* Choosing a packaging film material with low water vapour permeability to prevent moisture migration to or from the environment during storage is important. Other methods such as vacuum packaging or addition of chelators (e.g., EDTA), antioxidants, or oxygen scavengers in the packaging may also be used depending on the food composition.

The type of fatty acids in TAGs influences the reaction rate, and generally, *as the number of double bonds increases the reactivity increases*. For instance, oleic acid with 18 carbons and one double bond (18:1) oxidises ~100 times faster than stearic acid with 18 carbons and zero double bonds (18:0), whereas the rate of oxidation of α-linolenic acid with 18 carbons and three double bonds (18:3) can be more than 2000 times faster. Fish and plant oils are sensitive to oxidation and care needs to be taken during food formulation and storage. Furthermore, residual unsaturated fatty acids that remain in plant protein isolates (e.g., pea or soy protein) create oxidative stability problems for the long-term storage of products formulated with them. An increase in temperature, oxygen concentration, surface area (e.g., bulk oil *vs* emulsion or minced meat *vs* whole muscle), and light (UV and visible) increase the rate of lipid oxidation. Strategies that focus on minimising exposure of food to these factors generally improve oxidative stability.

Antioxidants can stop or slow down oxidation, but it is difficult to group them into one category, as they can delay oxidation by different mechanisms. Antioxidants exert their influence by controlling free radicals, pro-oxidants, or oxidation intermediates. Free-radical scavengers slow oxidation by reacting faster and scavenging (removing) free radicals resulting in inhibition of the initiation and propagation reactions. They donate hydrogen to peroxyl radicals forming low energy scavenger-radicals that are not efficient in propagating oxidation. Control of pro-oxidant metal ions (e.g., Fe, Cu), singlet O_2, or lipoxygenases is achieved with the use of chelators to remove metal cations (e.g., EDTA), the addition of carotenoids to quench singlet O_2, or heat treatments to deactivate enzymes (e.g., blanching). Antioxidants can be natural (e.g., tocopherol, ascorbic acid, or carotenoids) or synthetic (e.g., butylated hydroxyanisole (BHA), butylated hydroxytoluene (BHT), or tertiary butylhydroquinone (TBHQ), Fig. 4.13). The choice of antioxidant depends on food composition, desired shelf life, and cost. Generally, the ideal antioxidant must not have any harmful physiological effects or contribute to an objectionable flavour, odour, or colour to the product. It should be effective at low concentrations, soluble in the

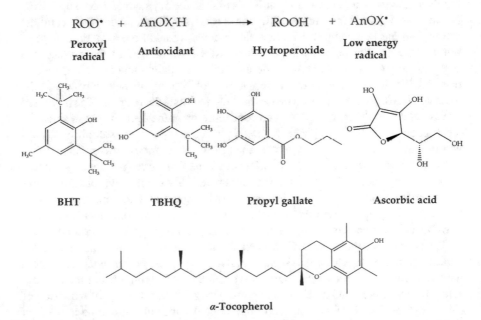

Fig. 4.13 General mechanism of action of free radical scavengers (top) and structures of some synthetic and natural antioxidants (bottom)

Table 4.3 Summary of factors influencing lipid oxidation

Parameter	Comments
Oxygen	Triplet and singlet O_2. Generally, higher O_2 concentration accelerates oxidation
Double bonds	Higher number of double bonds accelerates oxidation
Pro-oxidants	Transition metals, $^1O_2$, and lipoxygenase accelerate oxidation
Temperature	Higher temperatures accelerate oxidation
Surface area	Higher surface area facilitates oxidation, e.g., emulsion vs bulk oil
Water activity	Fast at very low a_w (< 0.2), low at intermediate a_w (0.2–0.5), and fast at high a_w (> 0.5)
Antioxidants	Delay $^3O_2$ oxidation. Limited effect on $^1O_2$ oxidation
Light	UV and visible light accelerate oxidation
Storage time	The longer we store the food, the more it will oxidise

lipid phase, and remain unaltered during processing. Cost, availability, and marketing issues (e.g., "natural") should also be considered when choosing the correct approach. Alternatives to the use of antioxidants could be the elimination of oxygen (e.g., packaging under nitrogen, vacuum or with an oxygen scavenger), elimination of the sensitive substrates (e.g., replacement of polyunsaturated oils with less unsaturated that are more stable), or by decreasing the rate of oxidation with physical methods (e.g., storage at low temperatures, storage in the dark, use of lipids that contain low levels of oxidation promoters or use of ingredients that are naturally rich in antioxidants). An overview of the factors influencing lipid oxidation is presented in Table 4.3.

Lipolysis is the hydrolysis of ester bonds in lipids by enzymatic action or heat and moisture, resulting in the liberation of free fatty acids leading to hydrolytic rancidity (Fig. 3.33). We should be careful not to confuse oxidative rancidity with hydrolytic rancidity, as the mechanisms of the reactions and products are entirely different. Lipolysis occurs in frying oil due to large amounts of water introduced from the food and the high temperatures. Enzymatic lipolysis has been described in Sect. 3.4.4.3. Lipolysis may have detrimental effects, such as in the destruction of essential fatty acids, free-radical damage to other compounds, including vitamins and proteins, but most notably in the development of off-flavour and off-odours in vegetables (e.g., beans and peas).

Fat Analysis

Fat analysis in foods proceeds by extracting the fat with a suitable organic solvent (e.g., petroleum ether), evaporating the solvent, and finally measuring the weight of the fat residue. These are known as *solvent extraction methods,* with the *Goldfish method* and *Soxhlet method* being the two most common representatives. Lipid characterisation also proceeds with a fat extraction step with a suitable solvent depending on food composition. Following extraction, many analytical techniques are available to characterise the properties of the extracted lipids depending on the purpose (Table 4.4).

Table 4.4 Common methods used in fat analysis with their primary purpose

Method	Purpose
Smoke, flash, and fire points	Identification of temperatures at which the sample begins to smoke (smoke point), ignites (flash point) or combusts (continuous burning) (fire point). Higher smoke point makes oil suitable for frying applications.
Melting point	Identification of the *temperature range* at which fat melts. Important for texture or storage considerations.
Iodine value	A measure of the degree of unsaturation. A higher number indicates more unsaturated lipid.
Saponification value	A measure of the average molecular weight of triacylglycerols in the sample. Smaller values indicate longer average fatty acid chain length.
Free fatty acids (FFA) and acid values	A measure of fat acidity reflecting the number of fatty acids hydrolysed from triacylglycerols during refining or storage. High FFA values reflect poor oil quality.
Solid fat content	It measures the amount of crystalline fat (solid fat) in a sample at a specific temperature.
Total polar compounds (TPC)	A measure of deep-frying oil quality. When a frying oil has TPC > ~27%, then it should be discarded.
Peroxide value	A measure of the amount of peroxide or hydroperoxide groups. Higher values indicate advanced lipid oxidation (see also next row).
Anisidine value and Totox value	Anisidine value determines the number of aldehydes, whereas the Totox value indicates the total oxidation of a sample using both the peroxide and anisidine values. Higher values indicate advanced oxidation.
Thiobarbituric acid test (TBA)	Measures malonaldehyde, which is a secondary product of lipid oxidation. Higher malonaldehyde concentration indicates advanced oxidation.

4.5 Fat Crystallisation

4.5.1 Crystal Formation

Crystal is a solid in which the atoms, molecules or ions are arranged in a repeating pattern that extends in all three dimensions. In amorphous solids, also called *glassy solids*, atoms or molecules have a random orientation in space. The repeating pattern of crystals is termed *unit cell*. Most crystals are *symmetric*, which means they are invariant (do not change) when subjected to certain transformations (e.g., rotation). Several types of symmetry exist, as for example symmetry about a point (centre of symmetry), line (axis of symmetry), or plane (plane of symmetry). Unit cells are classified into one of the *seven lattice systems* in terms of *rotational symmetry*.

The seven lattice systems are *cubic* (NaCl), *hexagonal* (ice), *tetragonal, rhombohedral, orthorhombic* (TAGs), *monoclinic* (most sugars), and *triclinic* (TAGs). TAGs can be molecularly organised to form crystals, and depending on composition and processing conditions, they may form crystals into one or more of the lattice systems mentioned above. TAG crystals are often formed during food processing and storage and influence food properties, mouthfeel, and physical stability. For instance, the presence and type of crystals may influence fracture properties, stickiness (when small fat crystals do not melt in the mouth), cooling sensation (crystal melting is an endothermic process creating a cooling effect during eating), or physical stability of some foods (e.g., formation and sedimentation of crystals in oils, phase separation or blooming of chocolate).

Crystals form through a process termed crystallisation. Crystallisation generally proceeds into two steps: *nucleation* followed by *crystal growth* (Fig. 4.14). During nucleation, the TAG molecules must become oriented into a fixed lattice and resist their tendency to redissolve. When a collision between molecules is unsuccessful, they redissolve, whereas a nucleus is formed when it is successful. A nucleus or *embryo* is a minute crystal that acts as a centre for further crystallisation (Fig. 4.14). The *critical radius* (r_c) represents the minimum size of a stable nucleus. Nuclei smaller than r_c dissolve, whereas larger than r_c continue to grow. Nucleation occurs from *solution* (i.e., sugar or salt solution) or *melt* (pure compound, e.g., only TAGs). Irrespectively the nucleation mode, a sufficient *thermodynamic force* needs to be generated to drive the process. Nucleation is most frequently achieved with either decrease in temperature (*supercooling*), or increase in solute concentration (*supersaturation*), or a suitable combination of both. The *cooling rate* plays a crucial role in the crystallisation and properties of TAGs. *Slow cooling rates* generate few big crystals with few impurities, whereas *fast cooling rates* generate many small crystals that grow fast and contain imperfections. Crystallisation occurs at higher temperatures with a slow cooling rate. In contrast, crystallisation occurs at lower temperatures (we need to decrease temperature more) with fast cooling rates. Control of cooling rate may be employed to manipulate the physical properties of TAGs and improve the stability and sensory properties of various foods (e.g., chocolate or margarine). After nuclei have been formed, the crystal continues to grow until the solution is not anymore supersaturated or all material has been crystallised (Fig. 4.14). Once the first nuclei have been formed, they act as the foundation for additional molecules to deposit (attach) on them, and the crystal increases in three

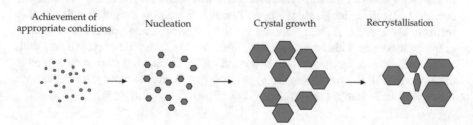

Fig. 4.14 Crystallisation process of food components

dimensions in successive layers. The pattern of growth usually resembles that of onion layers, with layers getting increasingly thinner due to the greater surface area of the growing crystal. *Impurities* (e.g., other food components present in formulation) may influence nucleation rate and crystal growth, usually by interfering with molecule deposition on the newly formed nucleus. In some cases, they may enhance nucleation acting as nucleation points. Impurities are very common in foods due to the complexity of formulations, and careful control of raw materials and processing conditions is required for reproducible products. Once crystals have been formed, *reorganisation* takes place during storage. This process, termed recrystallisation, is defined as the change of crystal size, number, shape, or orientation in time (Fig. 4.14). It occurs rapidly at storage temperatures close to the melting point, temperature fluctuations during transportation, or improper storage conditions. Recrystallisation is influenced by the composition of the product and the initial crystal size distribution (i.e., cooling rate during formation). There are five types of recrystallisation: *isomass rounding* (changes in shape during storage), *Ostwald ripening* (large crystals grow, small disappear), *accretion* (crystals that are close together grow as one (stick together)), *melt-refreeze* (due to temperature fluctuations during storage), and *polymorphic transitions* (changes in polymorphism during storage, see below). Accretion and melt-refreeze are the two most important recrystallisation mechanisms for most foods whereas polymorphic transitions are important in TAG-rich products only.

4.5.2   TAG Conformation and Polymorphism

The shape of TAGs in the solid phase is usually compared to either a tuning fork or a *chair* (Fig. 4.15). When TAGs crystallise, they align side-by-side to form a unit cell in a specific manner depending on their chemical properties. A *double chain length* structure forms when the chemical properties of the three fatty acids are the same or very similar. A *triple chain length* structure forms when the chemical properties of one or two of the three fatty acids are different from each other (Fig. 4.16).

The way TAGs associate in the crystal has a profound effect on the physical properties of fats. The most notable property is polymorphism, which is the occurrence of several different crystal forms originating from the same TAG. Polymorphic forms are crystals of the same chemical composition that differ in crystal structure but yield identical liquid phases upon melting. The most common polymorphic forms of TAGs are α, β', and β and they differ in their melting points and crystallographic properties. Other polymorphic forms are also possible (e.g., γ or α') with associated nuances in their physical properties. TAG stacking in α-polymorphs is usually vertical tuning forks, in β'-polymorphs is tilted tuning forks, and in β-polymorphs is stacked chairs (Fig. 4.17).

a.
sn-2
sn-1
sn-3
Tuning fork
b.
sn-2
sn-1
sn-3
sn-2
sn-3
sn-1
Chair

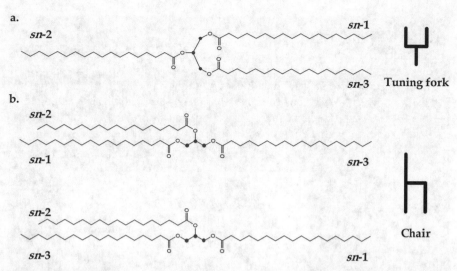

Fig. 4.15 (**a**) Tuning fork, and (**b**) chair conformations of TAGs. The red dots indicate the carbons of glycerol

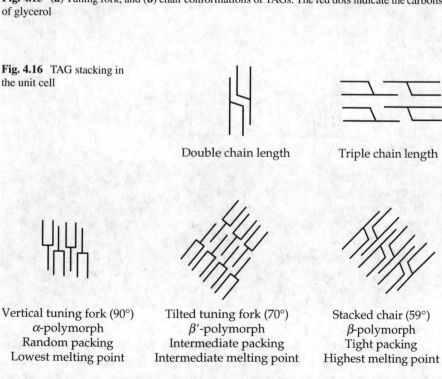

Fig. 4.16 TAG stacking in the unit cell

Double chain length
Triple chain length

Vertical tuning fork (90°)
α-polymorph
Random packing
Lowest melting point
Tilted tuning fork (70°)
β′-polymorph
Intermediate packing
Intermediate melting point
Stacked chair (59°)
β-polymorph
Tight packing
Highest melting point

Fig. 4.17 TAG stacking in different polymorphic forms of fats

Fat Crystal Hierarchy

Bulk fats have different levels of structure depending on the *length-scale* of observation (how deep we zoom in the structure). Molecularly, TAGs associate to form a *lamella* and stacking of multiple lamellas form *crystalline domains*, which in turn they form *nanoplatelets*. The nanoplatelets assemble into *spherulites* that, with further aggregation, finally form the *bulk* fat (Figure below redrawn with permissions from Tang and Marangoni, 2006, *Adv. Colloid Interface Sci.*128–130, 257–265).

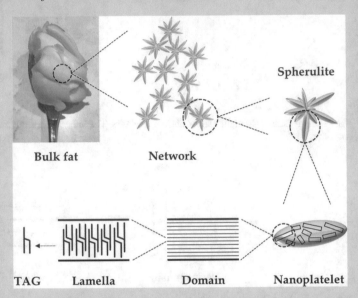

It should be emphasised that important technological properties of fats such as spreading, melting, or sensory properties depend on a combination of properties at the different structural levels. However, structures with size between 1 and 200 µm and the interactions between each other are primarily responsible for the technological properties of fats. As a result, the objective of the technologist would be to modify the crystal properties of fats at this length scale to improve or modify the physical properties or shelf life stability of the product.

In most cases, upon cooling a liquid fat, α-crystals form first because nucleation is easier in the α-form. In fats of relatively homogeneous composition, the α-form is short-lived and is transformed into β' that generally has a longer lifetime. At certain conditions (e.g., constant temperature), β'-crystals may transform to β-form. The irreversible process of going from the less stable (α) to the more stable form (β) depends on time and temperature and is termed polymorphic transition (or *transformation*) of fats. The transition rate also depends on TAG homogeneity. Homogeneous

TAGs change rapidly into the stable β-form. Fats with random TAG distribution may retain β'-form indefinitely. Processing factors such as formulation, cooling rate, or shear forces may affect the number and type of crystals formed. Since fats are complex mixtures of TAGs, the different polymorphic forms usually coexist. Polymorphic transitions during storage may change fat properties and influence quality, and therefore, formulations and processing parameters should be carefully controlled.

TAG stacking in crystals and possible polymorphic transitions bring about remarkable differences in thermal, morphological, and optical properties. Most importantly, the melting point (T_m), which is a crucial characteristic of fats increases as (low) $\alpha < \beta' < \beta$ (high). The properties of some products, such as chocolate or shortenings, are susceptible to the correct polymorphic form and only when the correct crystal is formed the product obtains its desirable characteristics. For example, cocoa butter has several crystal polymorphs and good chocolate can only be made from β-crystals. Tempering is a precise time-temperature process to ensure the formation of cocoa butter in the right crystal polymorph. All the undesirable polymorphic forms melt during tempering, and only certain β-crystals that have the optimum melting characteristics are retained. During storage at inappropriate temperatures or after incorrect tempering, fat moves from the centre of chocolate to the surface and crystallises. This is known as blooming (a white "mouldy" appearance at the surface) and is a major reason for product failure even though it is not a health hazard.

Shortening is a fat that is used in bakery products, and its polymorphic structure is responsible for the main quality characteristics. For example, in aerated bakery products (e.g., cakes or muffins), β'-crystals result in many small air bubbles, whereas β-crystals in a few large air bubbles. Consequently, β'-crystal shortenings are preferred as they help to form small air bubbles in batters that result in good volume, texture, and tenderness of baked goods. In addition, the melting point of shortening is critical for the mouth-melting effect in some bakery products. In margarine manufacturing, β'-crystals are important for smooth texture and spreading that is achieved with β' stable fats such as hydrogenated cottonseed oil.

Margarine is a water-in-oil (w/o) emulsion in which water is dispersed as droplets in a continuous fat phase. The lipid phase is usually a blend consisting of different fats and oils. The ratio and type of fats and oils in the fat blend are decisive to achieve margarine with the desired characteristics (e.g., spreading). Emulsifiers (lecithin, mono-, di-glycerides), flavours, colours, and antioxidants may also be added to achieve the desired characteristics. In margarine manufacturing, β' polymorphs are usually preferred as they are smaller, giving the product the appropriate spreading and textural properties. In contrast, β polymorphs form large crystals at a high solids content, resulting in brittle, hard margarine, while at a low solids content, the product becomes oily. Spreadability and mouth melting are two textural properties of margarine that depend on solid fat content and its crystalline structure. Good spreadability requires that the product retains its plasticity from the refrigerator to room temperatures (4–22 °C). However, desirable mouth-melting requires rapid melting at mouth temperature (35–37 °C) for prompt flavour release. It is

Table 4.5 Summary of factors influencing melting point (T_m) of TAGs. An interplay between all these factors influences the actual melting point

Low T_m	High T_m
Short-chain length	Long-chain length
Unsaturated	Saturated
Branched-chain	Straight-chain
cis-	trans-
Non-conjugated	Conjugated
α-polymorph	β-polymorph

evident that these two properties are in conflict with each other, and an in-depth understanding of crystal structure is required to fabricate margarine with desirable textural characteristics.

Apart from the crystal structure, chain length and number of double bonds in fatty acids also influence T_m. Generally, the greater the chain length the higher the T_m, while the increase in double bonds decreases T_m. Straight-chain fatty acids have higher T_m than branched fatty acids. Double bond configuration is also important as a *trans* double bond allows the formation of an almost straight-chain, causing a smaller difference in melting properties than a single *cis* bond compared to its saturated counterpart. Furthermore, the position of the double bonds in the chain is important as two conjugated *cis* double bonds ($-CH = CH-CH = CH-$) can give a nearly straight-chain than two non-conjugated ones ($-CH = CH-CH_2-CH = CH-$) (Table 4.5).

4.5.3 Other Properties of Fats

Natural fats are always mixtures of several different triglycerides. They, therefore, have a *melting range*, meaning that they melt over a range of temperatures rather than at a specific temperature, as ice at 0 °C. Food fats are rarely 100% solid at processing temperatures meaning that a certain proportion is always liquid at a specific temperature. The ratio of solid to liquid fat at a given temperature is termed solid fat content (SFC) and is important for the functional properties of fats (Fig. 4.18). SFC is a function of temperature, temperature history, time, and TAG composition. It influences the appearance and stability of products stored at refrigeration temperatures (e.g., salad dressings, spreading of margarine or butter etc.), the mouth-melting profile of chocolates, the texture of baked products, and ease of processing.

Chemical manipulation of TAG composition is also possible with the aim to change their physical properties. Hydrogenation is the addition of hydrogen to double bonds to convert liquid oils to semisolid fats (e.g., margarine, shortenings), change the crystallisation behaviour, or improve the oxidative stability of oils. It is usually carried out with H_2 and Ni as a catalyst (Fig. 4.19). Partial hydrogenation may result in double bonds in the *trans* configuration that causes concerns, as *trans*

Fig. 4.18 Solid fat content changes as a function of temperature. Hysteresis is also observed between heating and cooling

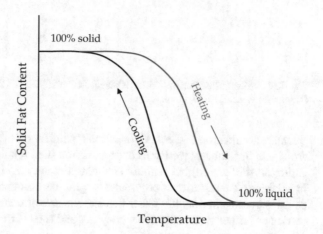

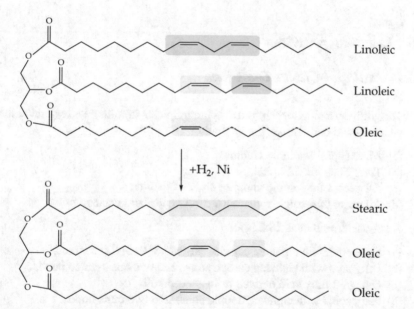

Fig. 4.19 Partial hydrogenation of TAGs. The double bonds in blue have been saturated with the addition of hydrogen (green). As hydrogenation is partial, two bonds have remained unaffected (yellow)

fatty acids have been associated with an increased risk of cardiovascular disease. However, the modern food industry moves towards alternative technologies such as complete hydrogenation, interesterification or other modern technologies (tropical fat fractionation, modified hydrogenation processes, crystallisation, and emulsification strategies) that do not result in the formation of *trans* fatty acids.

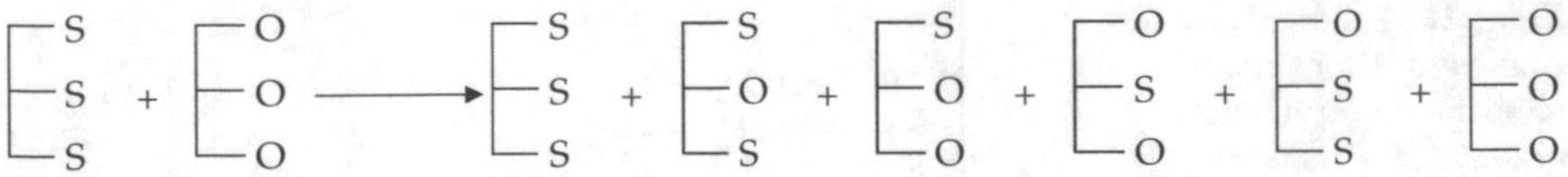

$$\begin{bmatrix} S \\ S \\ S \end{bmatrix} + \begin{bmatrix} O \\ O \\ O \end{bmatrix} \longrightarrow \begin{bmatrix} S \\ S \\ S \end{bmatrix} + \begin{bmatrix} S \\ O \\ S \end{bmatrix} + \begin{bmatrix} S \\ O \\ O \end{bmatrix} + \begin{bmatrix} O \\ S \\ O \end{bmatrix} + \begin{bmatrix} O \\ S \\ S \end{bmatrix} + \begin{bmatrix} O \\ O \\ O \end{bmatrix}$$

Fig. 4.20 Interesterification of two TAGs results in a mixture of TAGs with random positioning of fatty acids. S: stearic acid, O: oleic acid

Interesterification is the randomisation ("shuffling") of fatty acids in triacylglyc-erols with the same objective as hydrogenation (i.e., manipulation of melting point and/or crystallisation) but without changing the fatty acid composition (Fig. 4.20). In addition, interesterification creates negligible amounts *trans* fatty acids com-pared to hydrogenation. Interesterification consists of complex reactions that can be carried out either with catalysts or enzymes and is used to manufacture shortenings, margarine, or confectionery oils.

4.6 Learning Activities

4.6.1 Multiple-Choice Questions

1) Triacylglycerols, phospholipids, sphingolipids, steroids, waxes, fat-soluble vitamins, and carotenoids are all lipids

   (a) All except fat-soluble vitamins
   (b) True. They are all lipids
   (c) All except fat-soluble vitamins and carotenoids
   (d) All except fat-soluble vitamins, carotenoids, and waxes

2) A fatty acid written as 18:2 is a:

   (a) Fatty acid with eighteen calories (instead of 9) and zero trans.
   (b) Fatty acid with eighteen carbon atoms and two unsaturated bonds.
   (c) Omega-3 fatty acid derived from avocado oil.
   (d) Fatty acids with eighteen double bonds and two trans bonds.

3) A characteristic of trans- isomers is that:

   (a) Fatty acids that can be transformed to glycerol.
   (b) Can be transformed to polyunsaturated depending on the processing conditions.
   (c) Hydrogens attached to carbon atoms are on the same side of the dou-ble bond.
   (d) Hydrogens attached to carbon atoms are on either side of the double bond.

4) Triacylglycerols are formed by esterification of.

   (a) Glycerol and three fatty acids.
   (b) Glycerol two unsaturated fatty acid and one carotenoid.
   (c) Glycerol and three monosaccharides.
   (d) Glycerol and three amino acids.

5) Stereospecific numbering is:

   (a) A way to differentiate the degree of unsaturation of the fatty acid(s) on glycerol.
   (b) A way to differentiate the stereoisomerism of the fatty acid(s) on glycerol.
   (c) A way to differentiate the isomerism of the fatty acid(s) on glycerol.
   (d) A way to differentiate the positioning of the fatty acid(s) on glycerol.

6) The occurrence of several different crystal forms for the same lipid compound is called:

   (a) Isomerism.
   (b) Polyisomerism.
   (c) Polymorphism.
   (d) Crystal morphism.

7) Three most common different crystal forms of lipids are:

   (a) α, β', and β.
   (b) α, β, and γ.
   (c) α, β', and γ'.
   (d) α, γ, and α'.

8) The shape of triacylglycerols in the solid phase compares to a:

   (a) Tuning fork or chair.
   (b) Tuning fork or boat.
   (c) Plate and egg box.
   (d) Plate and spear.

9) Lipid oxidation has the following stages:

   (a) Initiation, propagation, termination
   (b) Activation, propagation, termination
   (c) Activation, preparation, termination, recovery
   (d) Activation, preparation, recovery

10) Hydrogenation is:

   (a) Addition of hydrogen to double bonds so we can create margarine.
   (b) Addition of hydrogen to double bonds so we can create butter.
   (c) Addition of hydrogen to single bonds so we can create margarine.
   (d) Addition of hydrogen to single bonds so we can create butter.

4.6.2   Short Answer Questions – Further Reading

1) Draw the structure of *cis*-linoleic acid (18:2 $\Delta^{9,12}$, both double bonds in *cis* configuration). Draw the structure of rumenic acid (18:2 $\Delta^{9,11}$) with *cis* at 9 and *trans* at 11.

2) **Online activity** Search online for "conjugated linoleic acids" and explore their structure and potential health benefits.

3) Draw the structure of the triglyceride *sn*-1-oleo-2-stearo-3-palmitin.

4) Draw the structures of phosphatidylinositol, phosphatidylethanolamine, phosphatidylcholine, and phosphatidylserine. Compare their structures and search online the term "membrane lipids".

5) A bakery makes vegan muffins using pea flour. Pea flour has about 1% fat that is rich in unsaturated fatty acids. After only 1 day at room temperature, the muffin gets a rancid off-flavour. Identify the cause of this defect. Discuss the chemistry of the reaction and the factors that influence its rate. What will you propose to stop the reaction and extend the shelf life of the muffin?

6) A chocolate manufacturer prepared a chocolate formulation with a melting point at ~26 °C, dull surface colour, and poor snap. What crystal polymorph has most likely been formed? How are you going to improve the physical properties of the chocolate? List the polymorphic forms of fats in order of increasing melting point.

7) **Online activity:** The main triglycerides of cocoa butter are POS, SOS, POP with P for palmitic, S for stearic, and O for oleic acid. Visit https://lipidlibrary. aocs.org and click on the Triglyceride Property Calculator (TPC). Form the triglycerides POS, SOS and POP and explore how their melting point changes with polymorphism.

8) **Online activity:** Visit https://lipidlibrary.aocs.org and explore the "Edible Oil Processing" tab.

9) A potato-crisp manufacturer uses vegetable oil for frying. Deep-fat frying causes hydrolysis and oxidation of the oil. Identify the products of these reactions. What strategies will you suggest to minimise the extent of oil hydrolysis?

10) **Advanced, Online activity:** Search online for "Diels-Alder reaction" and its relevance to frying oils.

11) **Further reading:** Find the article "Choe, E., Min, D. B. (2006) Mechanisms and factors for edible oil oxidation Comprehensive Reviews in Food Science and Food Safety, 5(4), 169-186".

   • Name and discuss the factors that affect edible oil oxidation.
   • Discuss the antioxidant mechanism of tocopherols and carotenoids in edible oils.

12) **Further reading:** Find the article: "Marangoni, A. G., van Duynhoven, J. P. M., Acevedo, N. C., Nicholson, R. A., & Patel, A. R. (2020). Advances in our understanding of the structure and functionality of edible fats and fat mimetics. Soft Matter, 16(2), 289-306."

- Discuss how the fat crystal network builds from TAG to bulk fat. Identify the different "length scales."
- Name four experimental techniques to assess the thickness of TAG nanoplatelets.
- Discuss the major strategies used to create fat mimetics.

4.6.3   Fill the Gaps

1) Fatty acids differ from one another in the _____ of carbon chain and the degree of _____.

2) The notation $\Delta^{x,y,\cdots}$ is used to indicate the positions of _____ bonds.

3) Naturally occurring fatty acids are always in ____ configuration.

4) Conjugated double bonds are separated by a _____ bond.

5) The prefix *sn* is used to differentiate the _____ of the fatty acid in a TAG.

6) Plant sterols are also known as _____.

7) Three types of lipid oxidation may be distinguished: _____, _____ and _____.

8) Peroxyl radicals react with another fatty acid to produce _____.

9) The melting point of fats increases as ___ < ___ < ___.

10) The process of a fat crystal change from the less stable to a more stable form is termed _____ _____.

11) To ensure the presence of correct polymorphic forms, chocolate needs to be _____.

12) The ratio of solid to liquid fat at a given temperature is termed _____ ____ ____.

13) Interesterification is carried out with the use of _____ or _____.

Chapter 5
Browning Reactions

Learning Objectives

After studying this chapter, you will be able to:

- Describe the mechanism of enzymatic browning
- Discuss deactivation strategies of enzymatic browning
- Discuss the reactions of non-enzymatic browning
- Distinguish between caramel flavour and caramel colour formation
- Describe Maillard reactions

5.1 Introduction

Food browning is divided into two types: *enzymatic* and *non-enzymatic*. The reaction needs specific enzymes to proceed in the first type, but they are not required in the second type. As we will see below, non-enzymatic browning can be further divided into reactions with distinct chemistry. The end-products of both reactions are dark brown pigments, but the chemical pathways to obtain them are quite diverse depending on the reactants. Browning reactions have significant implications for the food industry and are either desirable or undesirable. Enzymatic browning of fruit and vegetables results in up to 50% of production losses worldwide. Still, the reaction may benefit certain products such as tea or raisins, where enzymatic reactions provide their characteristic colour and flavour. Non-enzymatic browning is most of the time a beneficial reaction when the processing conditions are well controlled, although occasionally it may have negative implications. For instance, browning of cookies during baking is necessary, and it is not possible to think of appealing cookies without the appropriate amount of browning. In addition, non-enzymatic browning reactions may be used in new product development. For example, by modifying the roasting conditions of coffee beans, we can produce a broad range of coffee products even though we may start from the same raw material. As it becomes evident, depending on the specific food and purpose, we need to control the reactions and take advantage of their outcome for a particular technological reason. This chapter presents the chemistry behind the reactions and strategies to control them.

5.2   Enzymatic Browning

Enzymatic browning is one of the most important colour reactions affecting fruits, vegetables, and seafood. It is catalysed by the enzyme *polyphenol oxidase* (PPO), widely distributed in plants, fungi, arthropods, and mammals (Fig. 5.1). It is also referred to as phenoloxidase, phenolase, monophenol oxidase, diphenol oxidase, catechol oxidase or tyrosinase. The enzymes found in plants, animals, or fungal tissues are frequently different with respect to their primary structure, molecular weight, or substrate specificity. Still, they fundamentally catalyse the same reaction (isoenzymes).

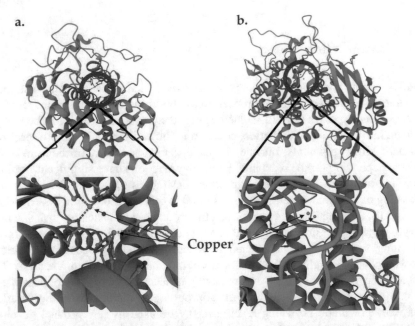

Fig. 5.1 Structures of polyphenol oxidases from (**a**) grapes, and (**b**) mushrooms. The zoom-in of the structures shows the copper ion required for activity. Although protein structures are different, they both have similar enzymatic activity. Figures were drawn in PyMOL based on PDB accession codes 2P3X and 5M6B

5.2.1   *Reactions and Substrates*

The reaction requires the *simultaneous presence of enzyme, substrate, and oxygen* to occur. In addition, PPO is a metalloenzyme, as it *contains copper* in its active site, and is required for its enzymatic activity. If one of these factors is absent, then the reaction cannot proceed.

Polyphenol oxidase catalyses two reactions: i) the hydroxylation of a mono-phenol to the *o*-position adjacent to an existing hydroxyl group of the phenolic substrate (*monophenol oxidase activity*), and ii) the oxidation of a di-phenol to *o*-benzoquinones (*di-phenol oxidase activity*) (Fig. 5.2). Both reactions utilise molecular oxygen as a co-substrate. The final products of the subsequent reactions are *melanins* that are dark brown water-insoluble compounds. Laccase is a type of copper-containing polyphenol oxidase with the unique ability of oxidising *p*-diphenols, in contrast to PPO that oxidises *o*-diphenols. They have low specificity and oxidise a wide variety of substrates. Laccases may be used to crosslink arabi-noxylans through the oxidation of ferulic acids and improve the baking performance of dough.

PPO substrates are phenolic compounds that contain an aromatic ring with one or more hydroxyl groups (e.g., tyrosine, catechol, chlorogenic acid, coumaric acid or catechins) (Fig. 5.3). Enzymatic browning does not occur in intact plant cells since phenolic compounds located in the vacuole are separated from PPO found in various plastids (e.g., chloroplasts). As a result, to observe browning, the tissues need to be ruptured either by physical damage (e.g., cutting or bruising) or by senes-cence (ageing) that brings together PPO and phenolic substrates. In crustaceans (e.g., shrimps), PPO is mainly located in the carapace (the "head"), whereas the

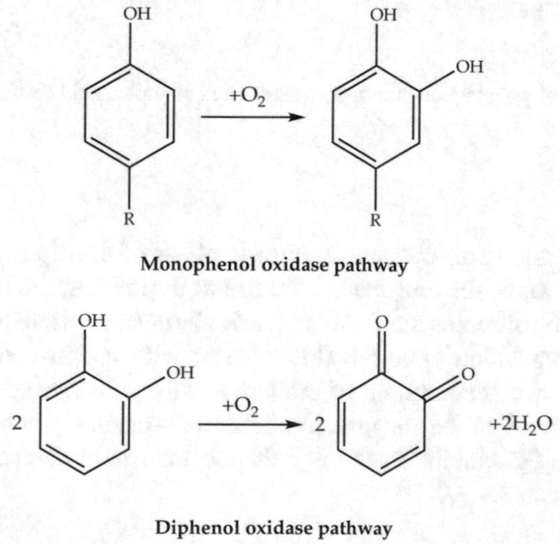

Monophenol oxidase pathway

Diphenol oxidase pathway

Fig. 5.2 Reactions catalysed by polyphenol oxidase. See text for description

a.   b.

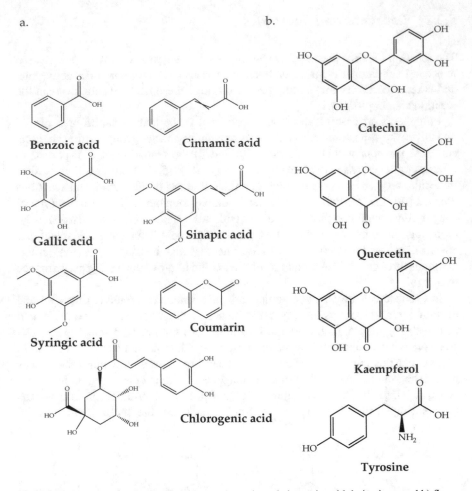

Fig. 5.3 Substrates of polyphenol oxidase a) benzoic and cinnamic acid derivatives, and b) flavonoids and catechins

substrates are released post-mortem from proteolysis of the digestive tract or other tissues. In mushrooms, the enzyme can be found throughout the tissues. The delicate structure of mushrooms makes them particularly susceptible to browning. The polyphenolic composition of commodities varies with species, cultivar, degree of ripening, environmental conditions of growth, and the substrate specificity of PPOs varies with the source of the enzyme. In addition to serving as substrates, phenolic compounds act as PPO inhibitors through the mechanism of competitive inhibition, as described in Sect. 3.4.2.4.

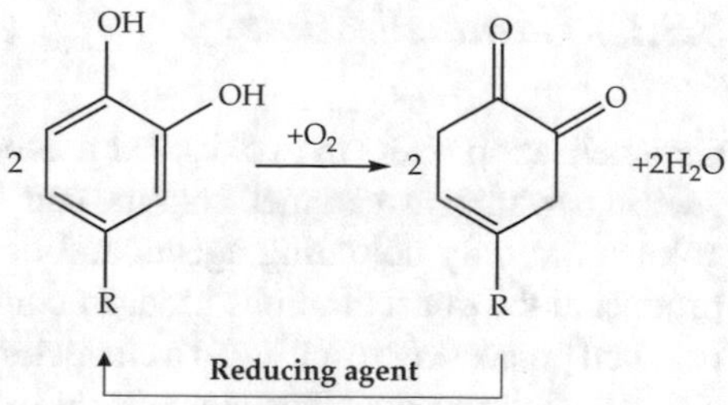

Fig. 5.4 Reducing agents reduce *o*-quinones to colourless *o*-diphenols

5.2.2 Deactivation Strategies

The strategies to deactivate PPO focus on eliminating from the reaction one or more of the essential components for its activity (i.e., oxygen, enzyme, copper, or substrate). There are generally five approaches that may be followed to control PPO activity:

1. *Inhibition targeted toward the substrate* includes *oxygen elimination* from the cut surface of fruits and vegetables. This can be achieved with vacuum packaging or immersion in water, syrup, or brine, depending on the product.
2. *Action on the enzyme* may include chelating agents that *remove copper* from the active site (e.g., EDTA, polycarboxylic acids) or *denaturation* of the enzyme. The latter is most frequently accomplished with heat treatments (e.g., blanching). *Lowering the pH* using various organic acids (e.g., citric, malic or tartaric acids) may also reduce activity, as the enzyme has an optimum pH of around 6.5.
3. *Low-temperature preservation*, such as refrigeration or freezing, reduce the rate of enzymatic reactions, but they do not eliminate activity.
4. *Dehydration* creates a low a_w environment that limits or stops the enzymatic activity. However, this is impractical as the time during dehydration is adequate for PPO to catalyse the browning reaction. As a result, an intermediate step is needed that targets the reaction products (see next).
5. *Inhibition targeted toward the reaction products* with chemical compounds that react with the products of PPO activity and inhibit the formation of colour. *Reducing agents* such as ascorbic acid and sulphites interfere with the reactions that lead to melanin formation, as they reduce pigment precursors (*o*-quinones) to colourless, less-reactive *o*-diphenols (Fig. 5.4). Ascorbic acid is commonly used in the fruit juice industry to maintain the colour of juices, purées, or during dehydration of fruit slices (e.g., banana chips or dried mango).

5.3 Non-enzymatic Browning

Caramelisation, ascorbic acid browning, and *Maillard reactions* are the three types of *non-enzymatic browning.* As the term implies, there are no enzymes involved in any of these reactions, and they all have different chemistry.

5.3.1 Caramelisation

Caramelisation refers to two distinct reactions that both involve sugars. One type of reaction results in caramel colours and the other in caramel flavours. Caramel colours are only colouring agents and do not influence the flavour profile of the product at the concentrations used. In contrast, caramel flavours are produced during food processing resulting in a characteristic flavour profile most frequently associated with the product. For example, the caramelisation of malt in some beer-making processes impacts the flavour profile of the beer. However, adding caramel colour to certain beers does not have any influence on their flavour. In another example, cookie-baking results in caramelisation of the top layer of dough that gives off volatiles characteristic of cookies. In contrast, adding caramel colour in the cookie dough to maintain colour uniformity does not influence the product's flavour. We should clarify that the food additives "caramel" (150a, 150b, 150c, 150d) always refer to a caramel colour.

5.3.1.1 Caramel Flavours

Caramelisation in the context of flavour creation is the controlled *thermal degradation* of sugars using acid or base. It generally requires high temperatures (>100 °C) and either a concentrated sugar solution or the absence of water. The reaction does not proceed rapidly at neutral pH (~ 7.0) and requires either acidic (pH < 3.0) or alkaline (pH > 9.0) conditions. The product is *caramels* that are complex polymeric mixtures of not well-defined compositions with different colour and flavour profiles. The caramelisation intensity and the exact final products vary depending on the type of sugar and reaction conditions. Fructose has the lowest caramelisation temperature (~110 °C) followed by galactose, glucose, and sucrose (~160 °C), maltose (~180 °C) and finally lactose (~203 °C). During caramelisation, volatile compounds are released, producing the characteristic *caramel aroma*. Furfural and maltol derivatives are the most important classes of these chemicals. Specifically, *hydroxymethyl furfural* (HMF), *hydroxy acetyl furan* (HAF), *hydroxy dimethyl furanone* (HDF), *maltol* and *hydroxy maltol* contribute to the aroma profile of many products, especially hard caramel, coffee, dried fruits, dark beers, and bakery products (Fig. 5.5). Additionally, *diacetyl* is also produced with a buttery flavour frequently found in fermented products, such as beer or yoghurt.

5.3.1.2 Caramel Colours

Caramel colours are among the most common colouring agents used in the food industry and produced *via* caramelisation reactions. Caramel colours are divided into four classes depending on the presence or absence of ammonium or sulfite compounds. In *Class I* or plain caramel (caustic caramel), ammonium or sulfite

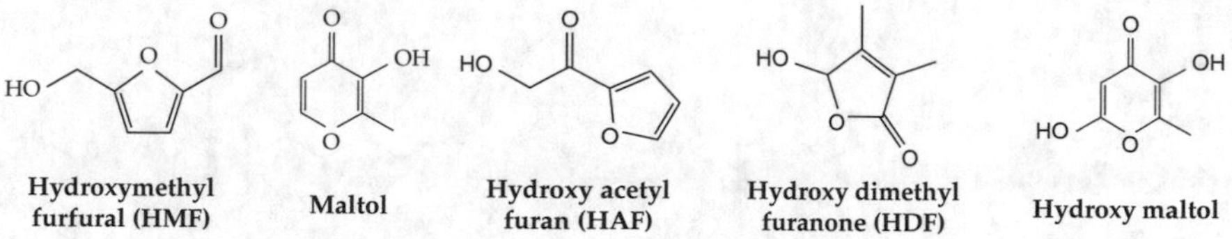

Structures of caramelisation products with a strong odour: Hydroxymethyl furfural (HMF), Maltol, Hydroxy acetyl furan (HAF), Hydroxy dimethyl furanone (HDF), Hydroxy maltol

Fig. 5.5 Structures of caramelisation products with a strong odour

Table 5.1 General characteristics of different classes of caramel colours

	Class I (150a) Plain caramel	Class II (150b) Caustic sulfite caramel	Class III (150c) Ammonia caramel	Class IV (150d) Sulfite ammonia caramel
Colour				
Sulphite	No	Yes	No	Yes
Ammonium	No	No	Yes	Yes
Charge	Slightly negative	Negative	Positive	Strongly negative
Stability in:				
Ethanol	+	+	−	−
Tannins	−	+	−	+
Acid	−	+	+	+
Applications	Distilled spirits, cereals, baked products, pet food	Tea, liqueurs (cognac, vermouth brandy)	Beer, sauces, gravies	Soft drinks carbonated beverages

compounds are not used during the reaction. In *Class II* (caustic sulfite caramel), the reaction is carried out in the presence of sulfite compounds but no ammonium compounds. *Class III* caramels (ammonia caramel) are produced in the presence of ammonium compounds but no sulfite compounds, and finally, *Class IV* (sulfite ammonia caramel) are produced in the presence of both sulfite and ammonium compounds (Table 5.1). The way caramel is produced influences its charge (positive or negative) and colour, making it a critical factor in the choice of caramel for a specific food application. For example, positively charged caramel colours may precipitate in the presence of negatively charged molecules in food, something that is undesirable. Caramel colour for soft drinks and beverages (e.g., cola or coffee flavoured drinks) is required to carry strong negative charges, whereas beer caramel colour should carry positive charges. In another example, caramel colours used in tea beverages need to have good stability in the presence of tannins, so they do not form complexes and precipitate.

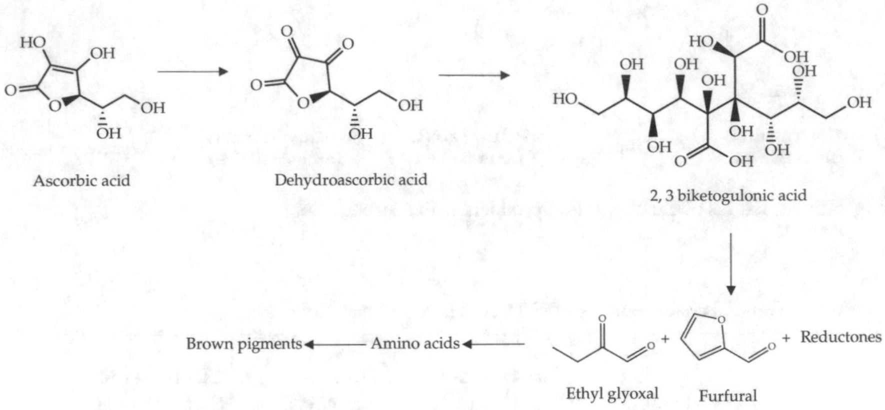

Fig. 5.6 Reaction pathway in ascorbic acid browning

5.3.2 Ascorbic Acid Browning

Ascorbic acid browning occurs in acidic foods such as fruit juices and some dehydrated fruits. In this type of browning, ascorbic acid (i.e., vitamin C) is lost lowering food's nutritional value. The reaction proceeds with the conversion of ascorbic acid to dehydroascorbic acid that further oxidises to 2, 3 biketogulonic acid (Fig. 5.6). The latter component is also prone to oxidation that results in the formation of furfural and ethyl glyoxal. These compounds may polymerise or react with amino acids through Strecker degradation (see Sect. 5.3.3.2) to form brown pigments. Decomposition of vitamin C takes place when in contact with oxygen. The brown colour only starts appearing once all vitamin C has been consumed in the reaction. It may occur at relatively low temperatures, and low pH and the presence of amino acids or proteins may facilitate colour production.

5.3.3 Maillard Reaction

Maillard reaction *is the reaction of a reducing sugar with an amino acid or protein*, and is of great importance in food manufacturing, as it may have either desirable (e.g., in bread, soy sauce, coffee, or chocolate) or undesirable results (e.g., in dehydrated foods such as potato flakes, egg and milk powders, or corn starch). Maillard reaction occurs every time a food is heated and a reducing sugar and amino acid are present in the formulation. As heat treatment is one of the most common food preservation techniques, it essentially signifies that Maillard reaction takes place nearly always and is perhaps *the most critical taste and flavour generating reaction in foods*. Some flavour changes are necessary and required in certain products (e.g., roasting coffee or cocoa beans). Occasionally, loss of essential amino acids may occur (e.g., lysine) in products already low in these amino acids, but this is not a significant problem when viewed from the perspective of a balanced diet. The reaction is particularly complex, and different reaction pathways are still being explored. Nevertheless, it can be divided into *three stages* depending on the colour and odour formation.

5.3.3.1 Early Maillard and Amadori Rearrangement

Stage 1 is the *colourless and odourless stage* of the reaction that is reversible. It starts with the reaction between the carbonyl group of a reducing sugar and the amino group of an amino acid yielding *N*-glycosylamine *via* a Schiff-base formation (Fig. 5.7). As it is a condensation reaction, i.e., water is produced, it is favoured at low water contents. Acidic environments protonate amines and cannot react, thus making it difficult for Maillard reaction to proceed. Generally, pentoses (e.g., arabinose) are more reactive than hexoses (e.g., glucose), while aldoses (e.g., glucose) are more reactive than ketoses (e.g., fructose). Monosaccharides are more reactive

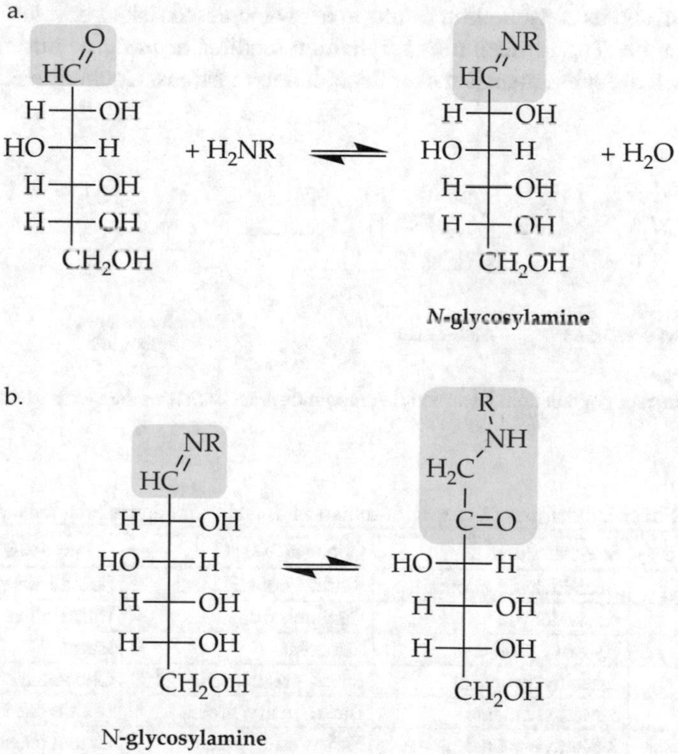

Fig. 5.7 Stage 1 of Maillard reactions (**a**) reaction of reducing sugar with amino acid, and (**b**) Amadori rearrangement forms Amadori compounds (ketosamines)

than reducing disaccharides (e.g., lactose, maltose), whereas non-reducing disaccharides (e.g., sucrose) do not react. Finally, strongly basic amino acids (e.g., lysine) are more reactive than other amino acids. Amadori rearrangement follows the first reaction where N-glycosylamine undergoes isomerisation to form ketosamines (Amadori compounds).

5.3.3.2 Advanced Maillard and Strecker Degradation

Stage 2 is where *colour and odour develop* and is the most diverse set of reactions that most frequently coincide. Amadori products are eventually *dehydrated* in a series of reactions that are not reversible. Deamination (loss of amine) and further isomerisation leads to several *dicarbonyl compounds* that are very reactive. It is only at this point that odour starts to develop, partly due to the appearance of HMF. Dicarbonyl compounds react with α-amino acids in a reaction known as Strecker degradation to produce amino-carbonyl compounds and Strecker aldehydes (Fig. 5.8). Strecker aldehydes have an intense smell and are important flavour compounds in many foods (Table 5.2).

Strecker degradation has also been implicated in acrylamide formation that is a carcinogen, and its consumption results in a low increased risk of developing cancer over a lifetime. The formation of acrylamide requires *asparagine* during the first reaction and the subsequent Strecker degradation reactions. Cooking methods such

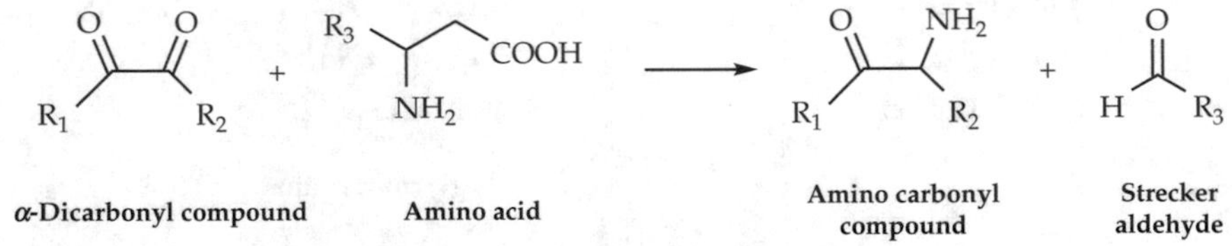

Fig. 5.8 Strecker degradation. Dicarbonyl compounds react with α-amino acids to form Strecker aldehydes

Table 5.2 Strecker aldehydes from specific amino acids and their odour characteristics

Amino acid	Strecker aldehyde	Odour at 100 °C	Odour at 180 °C
Cysteine	Acetaldehyde, propanal	Sulfide, cooked meat	Sulfide, smell of H_2S
Alanine	Acetaldehyde	Caramel, fruity	Burnt sugar
Threonine	Hydroxy propanal	Chocolate	Burnt
Valine	2-Methyl propanal	Straw, green, rye bread	Chocolate
Leucine	3-Methyl butanal	Bread, malty, green	Cream cheese
Isoleucine	2-Methyl butanal	Fruity, ester, green	Cream cheese
Glutamine	Pyrrolidone	Chocolate	Hard caramel

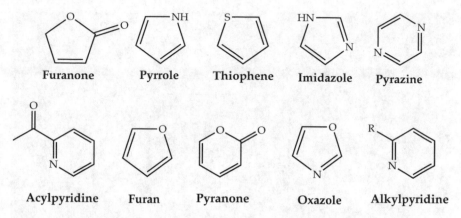

Fig. 5.9 Examples of heterocyclic compounds formed during Maillard reaction

as frying, baking, and roasting are likely to produce acrylamide, whereas boiling and steaming do not produce acrylamide. Acrylamide is formed in potato chips, fries, roasted coffee, cereal products, bread and bakery products, dried foods, chocolate products, biscuits. *Acrylamide formation cannot be avoided*, but it can be controlled with good manufacturing practices. Cyclisation reactions that result in ring formation may also occur during Stage 2 that contribute substantially to flavour generation. *Heterocyclic compounds* are ring-like structures, which contain atoms other than carbon in the ring (Fig. 5.9). Oxygen, nitrogen, and sulphur are primarily present in such molecules and play a key role in food flavour. These components may be present in minimal amounts yet, with very low detection thresholds, can significantly impact flavour profiles. Chocolate, caramel, and nutty flavours frequently occur because of the presence of heterocyclic compounds.

Acrylamide

The presence of acrylamide in foods is a very recent discovery (in the 2000s). Acrylamide in food is perhaps as old as the first-ever bread loaf that has ever been baked thousands of years ago! Acrylamide in high-doses, much higher than those encountered in food, is a carcinogen and a neurotoxin for humans. Acrylamide is not present in raw ingredients but is formed due to the high heat processing of carbohydrate-rich foods, especially during frying, baking, and roasting. French fries, bakery products, and coffee have the highest amounts of acrylamide because of the presence of reducing sugars and asparagine, whereas meat products contain very little acrylamide.

The formation of acrylamide is quite complex, and it involves several different mechanisms. The presence of 3-aminopropionamide (3-APA) seems to be a common characteristic in most of the reaction pathways. 3-APA may be produced in the reaction of asparagine with dicarbonyl compounds after decarboxylation of the formed Schiff base. Subsequent hydrolysis and deamination result in the formation of acrylamide.

5.3.3.3 Late Maillard and Melanoidin Formation

Stage 3 of the reaction involves polymerisation reactions that occur randomly between carbonyls that, in many instances, become attached to partly degraded proteins in the mixture. Although the polymerised pigments are initially water-soluble, as their size increases, their solubility decreases and precipitate. The final products are called melanoidins that have a particularly dark brown colour. An overview of all the stages that are involved in the Maillard reaction is presented in Fig. 5.10.

It should be noted that this idealised reaction scheme may be easy to be conceived when model reactions under laboratory conditions are discussed, e.g., the reaction of glucose with alanine in a test tube. However, in a food undergoing heat treatment, the complexity and diversity of the compounds that participate simultaneously makes distinguishing, understanding, and controlling each step virtually impossible. As a result, foods whose flavour profile is based on Maillard reactions (e.g., potato crisps or coffee) are susceptible to the quality and reproducibility of

Fig. 5.10 Overview of
Maillard reaction

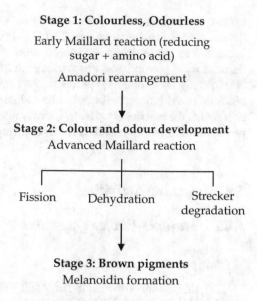

Stage 1: Colourless, Odourless

Early Maillard reaction (reducing
sugar + amino acid)

Amadori rearrangement

Stage 2: Colour and odour development
Advanced Maillard reaction

Fission   Dehydration   Strecker
degradation

Stage 3: Brown pigments
Melanoidin formation

raw materials (e.g., the genetic make-up of coffee beans) and processing conditions (e.g., time and temperature). For instance, reducing sugar content in potatoes vary with season, potato variety or storage time. As a result, understanding the exact chemistry of this reaction is relatively futile for technological purposes. In contrast, understanding the biology of raw materials, processing conditions, and quality control is of paramount importance to ensure reproducibility of the quality characteristics of these foods.

5.3.3.4   Control of Maillard Browning

Although precise control of the reaction at the molecular level may be an ineffective approach to use on the industrial scale, the general factors influencing Maillard reaction may be utilised to control browning. As with all other chemical reactions, an increase in *temperature* tends to speed up browning. The effect of *pH* is not as simple as that for temperature, as different reactions in the sequence are affected differently by changes in pH. Generally, acidic environments protonate amino acids and cannot react anymore with the reducing sugars. Browning occurs faster at a_w between 0.6 and 0.7 with moisture contents ~30%. This is normal for most chemical reactions, i.e., the higher the reactant concentrations, the faster the reaction rate. In addition, as the first step in Maillard browning is a condensation, lower moisture content facilitates the reaction. Below 30% moisture, there is a decreased rate of browning since reactants become trapped and cannot diffuse freely. If the reactants are not able to meet each other, then the reaction does not take place. As oxygen is not involved in the Maillard reaction it does not influence its rate. Different classes of compounds may act as accelerators or inhibitors, such as phosphates or various carboxylic acids and their salts, which may accelerate browning or result in more intense colour development. Tin (Sn) ions may slow browning to a certain extent because they cause strong reducing conditions. This is important when heating in tin cans, as in pasteurisation or sterilisation of various foods. Sulfur dioxide (SO_2) is frequently used in several products as a preservative, including beverages. It can block carbonyl groups as it reacts with reducing sugars and prevents the condensation step of the reaction. In addition, it may also react with products (e.g., HMF) and prevent further colour reactions from occurring.

5.4   Learning Activities

5.4.1   Multiple-Choice Questions

1) Polyphenol oxidase catalyses:

   (a) Enzymatic browning in fruits and vegetables
   (b) Non-enzymatic browning in fruits and vegetables
   (c) Enzymatic browning in fruits and vegetables that contain ascorbic acid
   (d) Enzymatic browning in fruits and vegetables that lack ascorbic acid

2) Substrates of polyphenol oxidase are:

   (a) Ascorbic acid and dehydroascorbic acid
   (b) Phenolics
   (c) Reducing sugars
   (d) Reducing sugars and amino acids

3) We can inhibit the action of polyphenol oxidase by

   (a) Blanching, elimination of N_2, copper removal
   (b) Blanching, elimination of O_2, zinc removal
   (c) Blanching, elimination of O_2, copper removal
   (d) Blanching, increase of O_2, zinc removal

4) Caramelisation requires:

   (a) reaction of non-reducing sugars with amino acids
   (b) acid or base, high temperatures and $3 < pH < 9$
   (c) low temperatures, free amino acids and reducing sugars
   (d) high temperatures, low pH (<3) and essential amino acids

5) Major products of caramelisation are

   (a) glucose derivatives and high and fructose derivatives at low temperatures
   (b) furanose and pyranose derivatives
   (c) reducing and non-reducing sugar derivatives
   (d) furfural and maltol derivatives

6) Maillard reaction is the reaction between

   (a) a reducing sugar and amino acid, peptide, or protein at high temperatures
   (b) a non-reducing sugar and amino acid, peptide, or protein at high temperatures
   (c) a reducing sugar and amino acid, peptide, or protein at low temperatures
   (d) a non-reducing sugar and amino acid, peptide, or protein at low temperatures

7) The Amadori rearrangement

   (a) Is the conversion of N-glycosylamines to 1-amino-1-deoxy-ketose in Maillard reaction
   (b) Is the conversion of N-glycosylamines to 1-amino-1-deoxy-ketose in ascorbic acid browning
   (c) Is the conversion of N-glycosylamines to 1-amino-1-deoxy-ketose in enzymatic browning
   (d) Is the conversion of N-glycosylamines to 1-amino-1-deoxy-ketose in Strecker degradation

8) Strecker degradation is the reaction between

   (a) dicarbonyl and ascorbic acid
   (b) dicarbonyl and non-reducing sugar
   (c) dicarbonyl and amino acid
   (d) dicarbonyl and reducing sugar

9) Strecker aldehydes are important because:

   (a) are responsible for important flavours in the Maillard reaction
   (b) are responsible for off- flavours in the Maillard reaction
   (c) are responsible for the creation of toxic compounds in the Maillard reaction
   (d) are responsible for the creation of brown pigments in the Maillard reaction

10) Acrylamide forms during the Maillard reaction and requires the presence of

   (a) Asparagine and a non-reducing sugar.
   (b) Asparagine and reducing sugar.
   (c) Arginine and reducing sugar.
   (d) Arginine and non-reducing sugar.

5.4.2 Short Answer Questions– Further Reading

1) Draw the monophenol and diphenol oxidation reactions of PPO.

2) Draw the structures of three benzoic and three cinnamic acid derivatives.

3) List and discuss the factors that affect enzymatic browning.

4) List monosaccharides and disaccharides in order of increasing colour formation in caramelisation.

5) A company makes a tea beverage containing tea extract. However, the colour between batches is inconsistent. Which caramel would you recommend and why?

6) List and discuss the factors that affect Maillard reaction. What is the difference between "melanins" and "melanoidins"?

7) A company manufactures potato crisps. The optimum level of reducing sugars in potatoes is 0.5 g / 100 g of potatoes. During quality control, the team detected unusually high levels of reducing sugars (1.5 g / 100 g). What is the effect of higher than the normal reducing sugars on colour, flavour, and acrylamide concentration of potato crisps?

8) **Online activity** Find online the terms "climacteric fruits" and "non-climacteric fruits". What is the difference between them? Which type is most likely to suffer from enzymatic browning?

9) Further **reading** Find the article "Lund, M. N., & Ray, C. A. (2017). Control of Maillard Reactions in Foods: Strategies and Chemical Mechanisms. *Journal of Agricultural and Food Chemistry*, 65(23), 4537-4552".

   • Discuss the impact of Maillard reactions on food product quality.
   • Discuss the impact of Maillard products on health.
   • Name all the strategies that we can used to control Maillard reaction.

10) **Further reading** Find the article "Yoruk, R., & Marshall, M. R. (2003). Physicochemical properties and function of plant polyphenol oxidase: a review. *Journal of Food Biochemistry*, 27(5), 361-422".

   • What is the importance of PPO in the food industry?
   • What is the optimum pH of apple, avocado, and pineapple PPO? What is the optimum temperature of apple, potato and strawberry PPO? Why do we have different optima even though the activity is the same?
   • Name the six groups of PPO activity inhibitors.

5.4.3   Fill the Gaps

1) Laccase has the unique ability of oxidising _____.

2) In crustaceans (e.g., shrimps), PPO is mostly located in the _____.

3) Reducing agents such as _____ acid and _____, interfere with enzymatic browning reactions.

4) Caramelisation in the context of flavour creation is the controlled _____ _____ of sugars.

5) _____ and _____ derivatives are the most important classes of chemicals with caramel aroma.

6) Caramel colours are divided depending on the presence or absence of _____ or _____ compounds.

7) Amadori compounds are chemically known as _____.

8) Dicarbonyl compounds react with _____ acids in a reaction known as _____ degradation.

9) Pigments formed in Maillard reaction are called _____. Those in enzymatic browning are called _____.

10) Formation of acrylamide requires the presence of _____ .

Chapter 6
Vitamins-Minerals

Learning Objectives

After studying this chapter, you will be able to:

- Recognise the structures of vitamins found in food
- Discuss vitamin stability during food processing and storage
- Recognise the minerals found in food
- Discuss the influence of food processing and storage on mineral bioavailability

6.1 Introduction

Vitamins and minerals may be essential nutrients but have only an ancillary role in creating and stabilising food structure. However, they may play a central role in organoleptic properties. For example, carotenoids may provide colour and pro-vitamin-A activity and sodium salty taste and the physiologically essential sodium cation. Nevertheless, vitamin degradation and mineral losses (e.g., Fe^{2+}) during food processing and storage is of concern. The vitamin and mineral losses in food are very complex and, in most cases, poorly understood. Irrespectively of how carefully we may process food, these nutrients are lost to a certain extent. Food processing aims to maximise vitamin and mineral retention by optimising processes that may degrade them. There are several causes of vitamin and mineral losses, starting with the inherent variation in their content in raw ingredients. For example, genetic characteristics of plants, stage of maturity, site of growth, soil, climate, agricultural practices, or animal diet may cause substantial differences in vitamin and mineral contents. Post-harvest changes resulting from the residual enzymatic activity may reduce vitamin content (e.g., lipoxygenases or ascorbic acid oxidase). Furthermore, continued plant metabolism (e.g., respiration) results in losses during storage. Food preparation steps such as trimming, washing, milling also result in vitamin and mineral losses. Blanching may result in thermal oxidation and losses due to leaching in the water and is important for water-soluble vitamins and minerals. Canning leads to variable losses depending on the food. Post-processing storage may also lead to vitamin deterioration, but it largely depends on food and is difficult to generalise but usually, the longer the storage, the greater the losses. Finally, additives and food composition, e.g., sulfites, nitrites, and oxidative food environment, affect vitamin losses. The present chapter outlines the essential chemical characteristics of vitamins and minerals and how they may behave in different processing conditions.

6.2 Fat-Soluble Vitamins

Vitamin A may be pre-formed vitamin A, including retinol, retinaldehyde and retinoic acid, found in animal-derived foods. It can also be pro-vitamin A that includes carotenoids (mainly β-carotene), which convert to retinal in the body (Fig. 6.1). Vitamin A is needed for good vision, skin, and for supporting the immune system.

β-Carotene has one-sixth of the vitamin activity of retinol. The total vitamin A content of the diet (from both animal and plant sources) is measured in µg of *retinol activity equivalents* (RAE). In the past, it was measured in international units (IU) but the unit is still widely present. An international unit (IU) is an arbitrary unit of measurement frequently used for fat-soluble vitamins and other pharmaceutical substances. As some vitamins may have a different structure but the same activity (e.g., retinol and β-carotene), IU may be used to compare the various forms with the same biological effect. For example, one IU of retinol equals 0.3 µg RAE, one IU of β-carotene from dietary supplements equals 0.15 µg RAE, whereas one IU of β-carotene from food equals 0.05 µg RAE. Sweet potato, liver, spinach, carrots are some sources that are rich in vitamin A or pro-vitamin A. Vitamin A degrades similar to oxidative degradation of unsaturated fatty acids. It is unstable to heat, light, acid, and oxygen, with up to 40% losses during thermal processing. Vitamin A

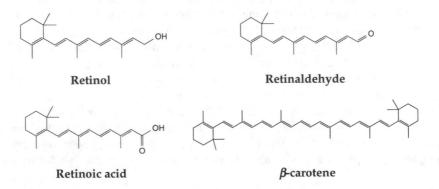

Retinol

Retinaldehyde

Retinoic acid

β-carotene

Fig. 6.1 Structures with vitamin A activity

Fig. 6.2 Structures with vitamin D activity

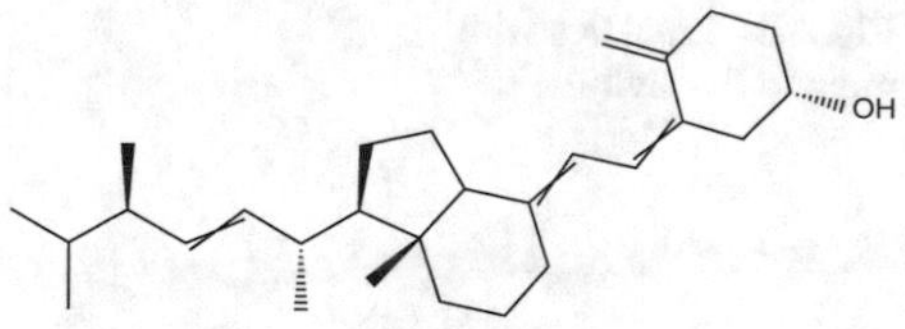

Ergocalciferol

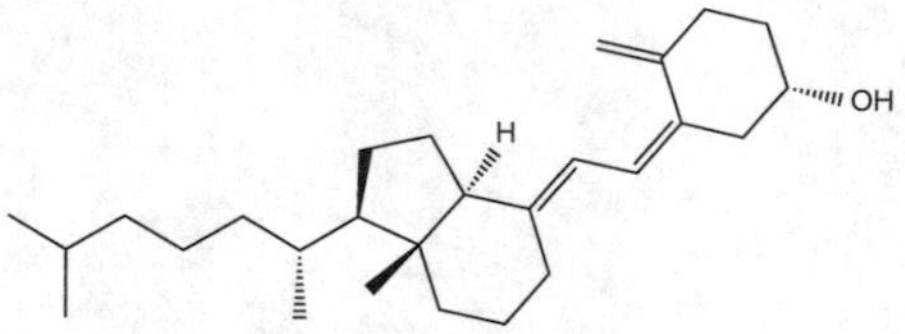

Cholecalciferol

degradation may be accelerated in dehydrated foods where oxidation generally proceeds faster.

The two most important forms of *vitamin D* are *vitamin D₂* (ergocalciferol) and *vitamin D₃* (cholecalciferol) (Fig. 6.2). Vitamin D promotes intestinal absorption of calcium and is involved in many other metabolic reactions. Vitamin D_2 is formed by the UV irradiation of the plant sterol ergosterol found in plants and fungi. Vitamin D_3 is synthesised in animals and humans after exposure to sunlight. Dietary vitamin D sources are practically insignificant compared to the amount prepared *via* the biosynthetic pathway (sunlight irradiation). Liver, beef, egg yolk, dairy products, some fish (e.g., salmon, tuna, and sardine) and fortified foods (e.g., milk, yoghurt, or margarine) are good sources of vitamin D. It is a stable vitamin during both heat processing and storage but is sensitive to light and oxygen. As a result, packaging materials of foods rich in vitamin D need low permeability to oxygen and light. Like vitamin A, vitamin D can also be measured in IU. One IU is equal to 0.025 μg of cholecalciferol or 1 μg of cholecalciferol is equal to 40 IU.

Vitamin E is a group of fat-soluble compounds consisting of four tocopherols (α-, β- γ-, and δ- tocopherol) and four tocotrienols (α-, β- γ-, and δ- tocotrienols) and protect cell membranes from oxidation. The main difference in their structures is that tocopherols have saturated side chains, whereas tocotrienols have three double bonds in their side chains (Fig. 6.3). It is found in plant oils and processed food rich in plant oils (e.g., salad dressings), fortified cereals, oatmeal, wheat germ, asparagus, tomatoes, and green leafy vegetables. It is sensitive to light, oxygen, and heat, especially in dehydrated or deep-fried foods. Processing and storage of foods may result in substantial tocopherol losses. However, anaerobic processing (e.g., canning) has little effect. Vitamin E frequently acts as a natural antioxidant by quenching singlet oxygen. 1 IU of vitamin E is equivalent to 0.67 mg of α-tocopherol.

Vitamin K is a family of naphthoquinones with or without terpenoid side chain (Fig. 6.4) and is needed for blood coagulation. It is found in plants, plant oils, fish oils, and meats. In humans, it is synthesised by the bacteria in the colon, where it is

Fig. 6.3 Structures with
vitamin E activity

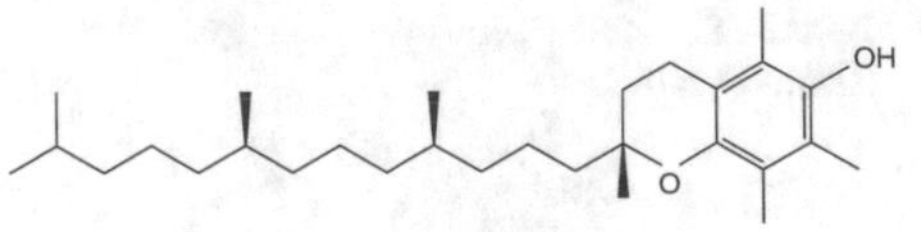

α-Tocopherol

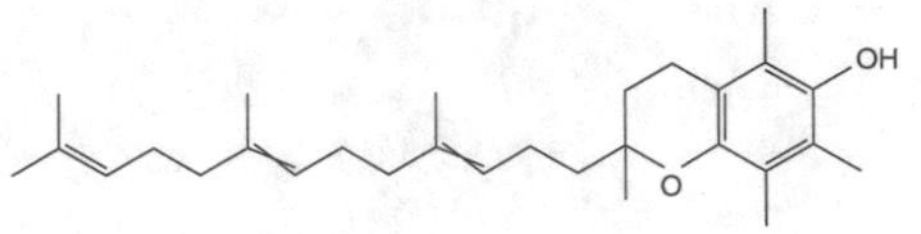

α-Tocotrienol

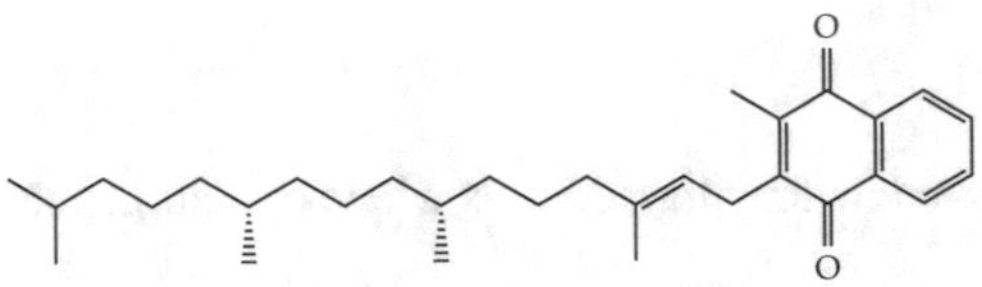

Phylloquinone 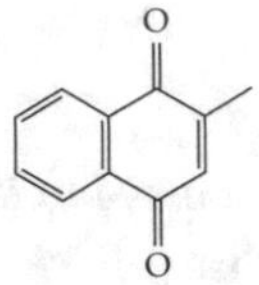   Menadione

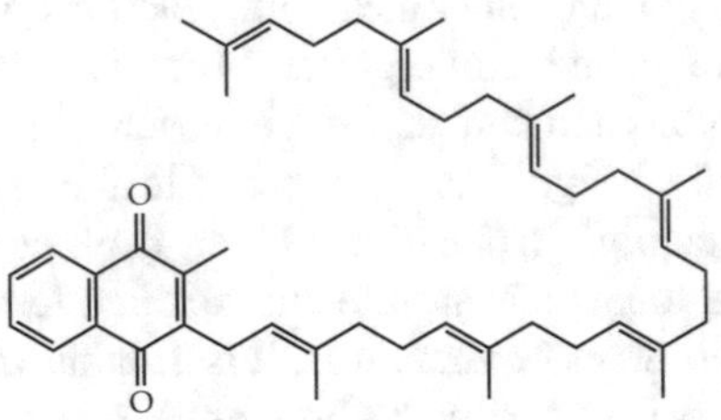

Menaquinone-7

Fig. 6.4 Structures with vitamin K activity

also absorbed (10% of needs). Good sources of vitamin K are dark green vegetables (e.g., spinach and cabbage leaves), cauliflower, peas, and cereals. It is very stable to heat and oxygen but not to light.

6.3   Water-Soluble Vitamins

Ascorbic acid (*vitamin C*) occurs in fruits (especially citrus fruits) and vegetables but not in animal tissues (Fig. 6.5). It is highly unstable to oxygen, light and heat, and it easily leaches out from tissues during processing. Ascorbic acid is oxidised easily to dehydro-L-ascorbic acid that retains vitamin C activity. However, this compound can be further oxidised to 2,3 biketogulonic acid that does not have biological activity (see Sect. 5.3.2). It is one of the most unstable vitamins, and there are

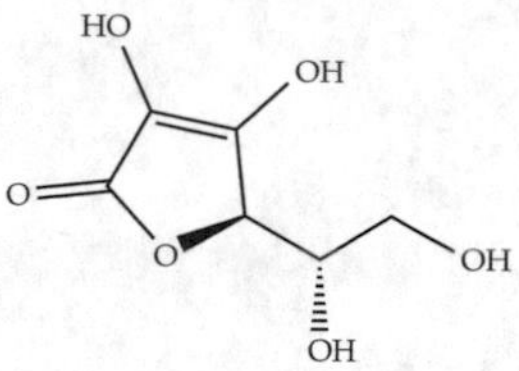

Fig. 6.5 Structure of ascorbic acid

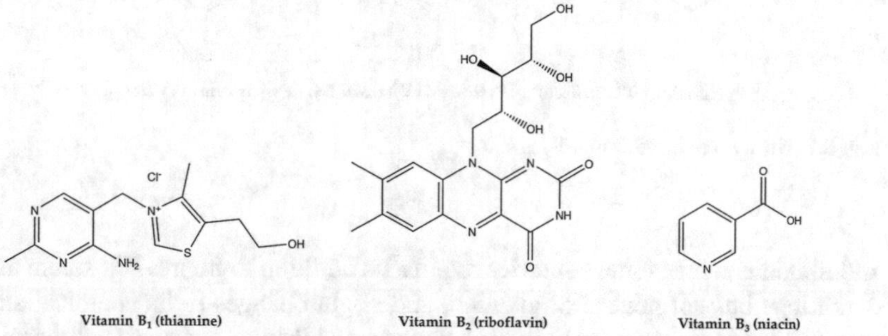

Fig. 6.6 Structures of vitamins B₁, B₂, and B₃

Vitamin B₁ (thiamine)   Vitamin B₂ (riboflavin)   Vitamin B₃ (niacin)

large losses during processing, storage and freezing (e.g., frozen fruits and vegetables), but it can be easily supplemented in fruit juices. Enzymes naturally found in tissues can also catalyse vitamin C decomposition, such as *ascorbic acid oxidase* or *peroxidase*. Deactivation of these enzymes is most frequently accomplished with blanching. Ascorbic acid can be used as a dough conditioner, antioxidant, an inhibitor of browning in fruits and vegetables, or colour stabiliser in meat products, just to name a few applications.

Vitamin B₁ (thiamine) is involved in the metabolism of carbohydrates, and food sources of thiamine include whole grains, legumes, and some meats, fish, and breakfast cereals (Fig. 6.6). Grain processing removes much of the thiamine content, and many countries enrich cereals and flours with thiamine. Thiamine is stable to light but among the least stable vitamins at neutral or alkaline pH, oxygen, and heat. This is important, as even simple food processing steps such as heating in water may result in significant losses either by decomposition or leaching. Vitamin B₁ has excellent stability in low a_w foods (e.g., breakfast cereals, bread), whereas tannins may interact and inactivate it. Nitrites found in processed meat products also inactivate thiamine. Losses in food may range from as little as ~10% in canned fruits and vegetables to up to ~60% in home-cooked meat. Regardless of the processing, ~20% of thiamine is generally lost.

Vitamin B₂ (riboflavin) is involved in energy production (Fig. 6.6). Food sources include eggs, green vegetables, milk and other dairy products, meat, mushrooms,

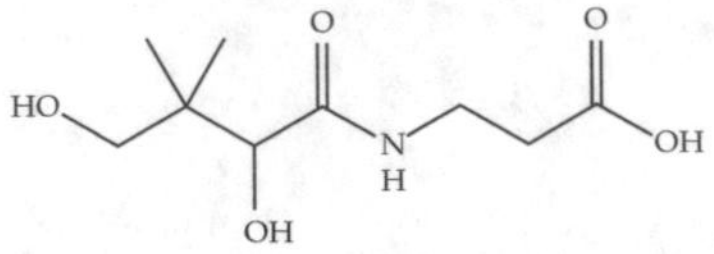

Pantothenic acid (Vitamin B5)

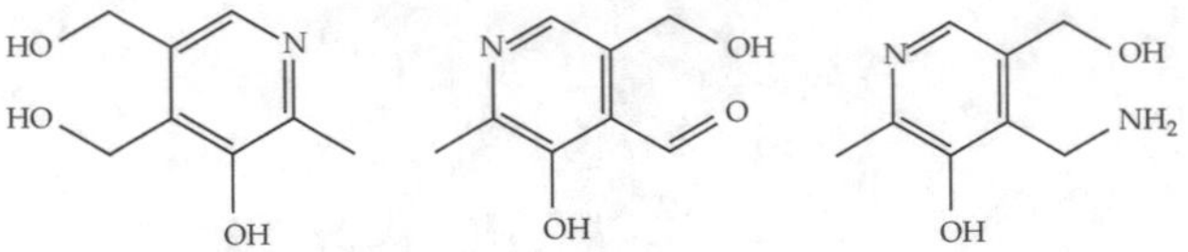

Pyridoxine (Vitamin B6)   Pyridoxal (Vitamin B6) Pyridoxamine (Vitamin B6)

Fig. 6.7 Structures of vitamins B_5 and B_6

and almonds while some countries require its addition to flours. It is stable at acidic conditions but not neutral or alkaline, and slight changes in its chemical structure result in loss of vitamin activity. It has good stability to conventional thermal processing, storage and oxygen but is very sensitive to light.

Vitamin B_3 (niacin) is the generic term for *nicotinic acid* and its derivatives involved in multiple metabolic activities (Fig. 6.6). Niacin is obtained from various whole and processed foods, with the highest contents found in fortified packaged foods, liver, tuna, salmon, and leafy vegetables. Some countries require its addition to flours similar to vitamin B_2. Niacin is the most stable vitamin, as it is not affected by light. Thermal losses do not occur under conditions relevant to food processing. Losses of ~15% may occur during blanching (leaching) or dripping on frozen meat thawing.

Vitamin B_5 (pantothenic acid) is found in both free and bound forms with 85% occurring bound as a part of coenzyme A (CoA) (Fig. 6.7). It is involved in fat metabolism and other important metabolic activities. It is widely available in meats, whole grains, potatoes, egg yolks, broccoli, mushrooms, avocados. It has good stability during food storage, especially in low a_w foods but is sensitive to heat and leaching during processing.

Vitamin B_6 refers to chemically similar compounds (pyridoxine, *pyridoxal, pyridoxamine*), which can be interconverted in biological systems (Fig. 6.7). Its active form is pyridoxal 5′-phosphate and it plays an important role in the metabolism of amino acids. Foods that contain large amounts of vitamin B_6 include fortified breakfast cereals, pork, turkey, beef, bananas, chickpeas, and potatoes. Pyridoxine has good stability to heat and low pH, but it is sensitive to light. Pyridoxamine is very sensitive to air, heat, or light and is practically destroyed during food processing. Pyridoxal is relatively stable and is the form that is used in food fortification. Fruit and vegetable canning results in ~25% losses, and wheat milling of up to 90% due to the separation of the bran.

Importance of Vitamins for the Food Technologist

Many consumers worry about the adequate consumption of vitamins from their food. It is also a popular misconception that processed food lacks or has insufficient vitamins. Modern food processing not only minimises the losses but, in some instances, enhances the concertation of vitamins in the food, with notable examples of the addition of vitamin B complex in breakfast cereals or vitamin D in fortified milk. In certain cases, modern technologies, such as *encapsulation*, minimise vitamin degradation during processing and storage and improve their absorption once in the gastrointestinal tract. Because of uncontrolled temperatures and inaccurate cooking methods, loss of vitamins during home cooking is more likely to occur than in industrial food preparation.

Regardless of their nutritional importance, food scientists rarely focus on vitamins. Vitamins do not constitute elements of food structuring and do not influence sensory properties, unlike minerals that may contribute to both structure creation (e.g., through calcium cross-linking) and taste (e.g., NaCl). Furthermore, vitamins rarely take part in reactions that influence food quality and have negligible technological significance. Some examples include ascorbic acid when used as a processing aid (e.g., gluten cross-linking in flours or colour stabiliser meat products) or involved in browning, the colour of β-carotene, or the antioxidant activity of tocopherols in some products. It is not an exaggeration to state that vitamins are technologically the least important of all molecules presented in this book!

Vitamin B_{12} (cobalamin) has a complex structure with cobalt (Co) chelated in the molecule (Fig. 6.8). It is important for the function of the nervous system and red blood cell formation. Various forms of cobalamin exist in which different groups are attached to the central Co atom, such as *cyanocobalamin, hydroxy-cobalamin, adenosyl-cobalamin* (major storage form in the liver), and *methyl-cobalamin* (mostly found in blood circulation). Meat, fish, milk, cheese, eggs, yeast extract, and fortified breakfast cereals are good sources of B_{12}. It is a stable vitamin and losses are negligible under most conditions of food processing.

Folate (folic acid) is made up of three distinct parts that all must be present for vitamin activity: *pteridine* (or pterin), *p-aminobenzoic acid* (PABA) and *L-glutamic acid* to form folate (Fig. 6.9). Folate is necessary for cell division and is required in amino acid metabolism. Although humans can synthesise all the parts of the vitamin, they do not have the enzyme needed for the coupling of the pterin molecule to PABA to form pteroic acid. Green vegetables, legumes and fortified grain products are good sources of folate. *Folic acid* is the synthetic form of folate used in supplements and food fortification. There are large differences in stability depending on the exact structure of folate. Naturally occurring folates are unstable and sensitive to oxygen, oxidative agents, heat, and light. However, folic acid is very stable in acidic environments and high-temperature processing.

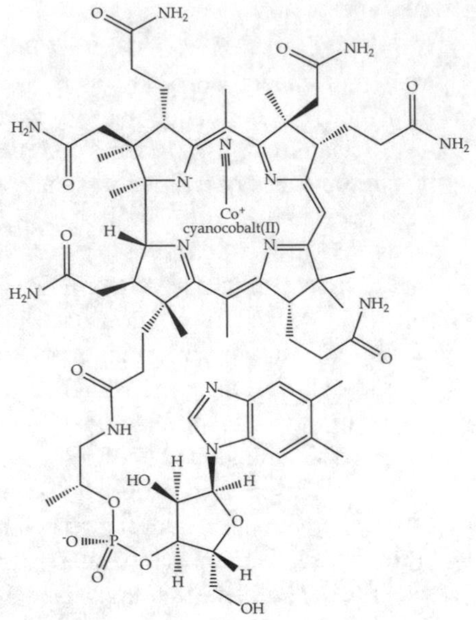

Fig. 6.8 Structure of cyanocobalamin (B_{12})

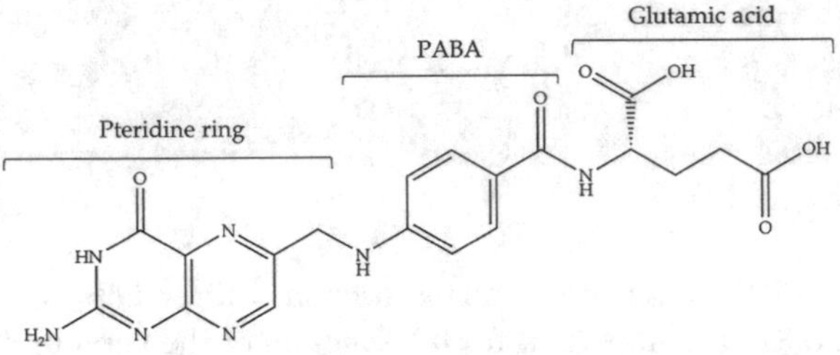

Fig. 6.9 Structures of folates (top) and biotin (bottom)

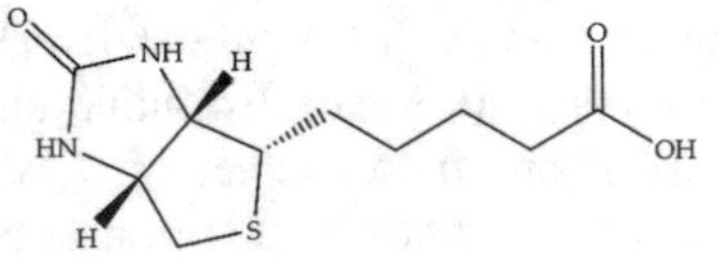

Biotin is involved in many metabolic processes related to all macronutrients (i.e., fats, carbohydrates, and amino acids), and it may be synthesised by intestinal bacteria (Fig. 6.9). Liver, soybeans, egg yolk, cereals, legumes, nuts are good sources of biotin. It is found combined with lysine (biocytin) or proteins, and proteolysis is required before its absorption. *Avidin*, a protein in egg white, binds biotin, making it unavailable, but denaturation of egg white during cooking improves availability. It is very stable to heat, light and oxygen, and is well retained in foods.

Table 6.1 Generalised stability of vitamins during processing and storage. U: unstable, S: stable. Stability may vary depending on food formulation, processing conditions and specific type of vitamin

Vitamin	Neutral pH	Acidic pH	Air or O_2	Light	Heat
A	S	U	U	U	U
D	S	S	U	U	U
E	S	S	U	U	U
K	S	U	S	U	S
B1	U	S	U	S	U
B2	S	S	S	U	U
B3	S	S	S	S	S
B5	S	U	S	S	U
B6	S	S	S	U	U
B12	S	S	U	U	S
Folate	U	U	U	U	U
C	U	S	U	U	U
Biotin	U	U	S	S	S

Vitamins are frequently added to foods to compensate for the losses occurring during processing or enhance them with a specific vitamin. *Restoration* refers to vitamin addition to restore their original concentration (e.g., vitamin C in orange juice). *Fortification* is the addition of nutrients to make food an excellent source of vitamin (e.g., breakfast cereals). *Enrichment* is the addition of specific amounts of vitamins lost during processing (e.g., flour enrichment with niacin, thiamine, riboflavin, and folic acid). Finally, *nutrification* is a generic term that encompasses the addition of nutrients to food (e.g., vitamin D addition in milk).

Some strategies may be followed to optimise vitamin retention in processed foods. The most important is the optimisation of thermal processing, where most of the losses generally occur, with lower intensity thermal processes being preferable when possible (e.g., HTST). Blanching should be carried out at lower temperatures with shorter exposure to hot water and a minimum water volume to minimise leaching. Prediction of losses with the determination of the rate of vitamin degradation under the processing conditions could be a valuable tool to optimise processing parameters. Finally, packaging using correct geometry in canning and use of materials with minimum oxygen and light permeability should be preferred. Table 6.1 presents a summary of vitamin stability under typical processing and storage conditions. It should be emphasised that this table should be used as a guideline, as stability highly depends on food formulation and storage conditions. For instance, the same vitamin may have additional oxidative stability in foods rich in antioxidants even if when they are processed under the same conditions.

6.4   Minerals

A *mineral* is a chemical element that is an essential nutrient, as it is needed to sustain life and promote growth. Minerals cannot be synthesised and must be obtained from the diet, but only small amounts are needed. The inorganic residue from the incineration (burning) of food is termed *ash* and contains minerals. Minerals are classified into *macro-minerals* (Ca, P, Na, K, Mg, Cl, S) and *trace elements* (Cu, I, Co, Fe, Mn, Se, Zn, Mo, F). Na, K, F, and Cl are present in foods as free ions with high bioavailability and rare nutritional deficiencies. The most common dietary deficiencies occur with Fe, Zn, Ca, and I. Mg, P, and Se are present in foods but bound as complexes, while both Cu and Mn have low bioavailability. Other elements may be found in foods but without nutritional value classed as *toxic metals*, posing a significant health risk such as Pb, Hg, As, Cd, or Cr. For example, lead and mercury may be found in high levels in seafood (e.g., mussels) and arsenic in rice.

The availability and reactivity of minerals *depend on their water solubility*. If the mineral is insoluble in food, then it is not absorbed. Bioavailability ranges from 1% for some forms of iron to 90% for sodium and potassium. Minerals may form complexes with compounds termed ligands. In foods, ligands include proteins, carbohydrates, phospholipids, and organic acids. Ligands and metals form complexes termed *chelates* (Fig. 6.10). Chelate formation stabilises the metal and may either improve or reduce its adsorption.

Organic acids generally enhance bioavailability (e.g., ascorbic, citric, and lactic acids). An organic acid of interest is phytic acid, the principal storage form of phosphorus in bran and seeds and is present in many legumes, cereals and grains. Phytic acid and phytate have a strong binding affinity to the dietary minerals, calcium, iron, and zinc, inhibiting their absorption that may lead to mineral deficiencies (e.g., in vegan diets or developing countries) (Fig. 6.11). Phytic acid, primarily as phytate in the form of phytin, is found within the hulls of seeds, including nuts, grains, and pulses.

Phytochemicals like polyphenols and tannins also influence the binding of minerals. They form insoluble precipitates and are far less absorbable in the intestines. Typical examples that restrict mineral absorption are polyphenolic compounds from tea, coffee, raisins, or sorghum. For example, a common practice of tea or coffee consumption immediately after a meal interferes with mineral absorption, including iron, zinc, and calcium that in the long-term may lead to mineral deficiencies.

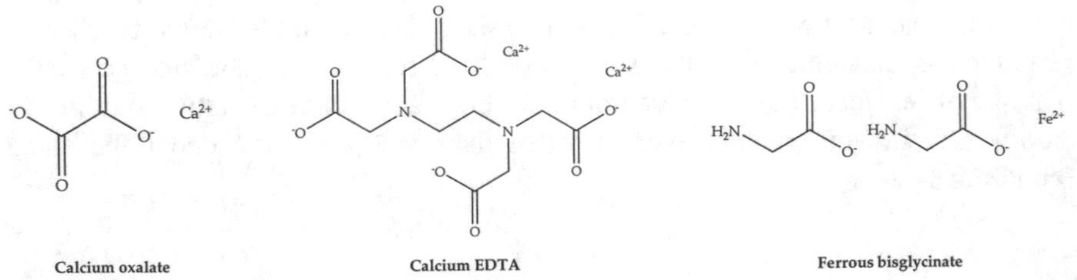

Fig. 6.10 Common chelates found in food

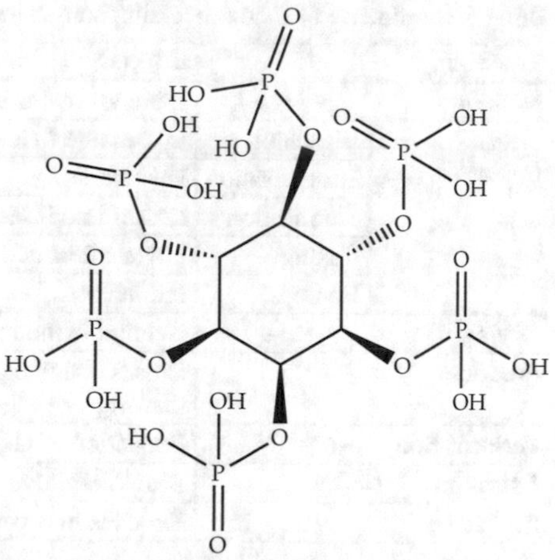

Fig. 6.11 Structure of phytic acid

Mineral and Vitamin Analysis

Ash content is measured after incineration of the sample at a high temperature (525 °C) in a muffle furnace. At such high temperatures, all the organic material is burnt, and the inorganic residue that is left contains the minerals. The ash may be further analysed to identify and quantify specific minerals of interest using atomic absorption spectroscopy (AAS) or inductively coupled plasma mass spectrometry (ICP-MS). Vitamin analysis in foods is quite complex. Methods usually have several steps and use specialised chromatographic protocols that need to be tailored for the specific vitamin and food under consideration.

6.4.1 Effect of Processing on Mineral Bioavailability

In minerals, *bioavailability is more important than content*, as high content does not guarantee absorption. Processing methods should influence solubility or destroy the inhibitory effect of phytic acid or tannins. Minerals are not destroyed by the factors that usually influence vitamins, i.e., heat, light, or pH, but they may change their form. For example, iron changes from Fe^{2+} (ferrous) to Fe^{3+} (ferric), and its bioavailability depends on the added form and interactions with other food constituents. Ascorbic acid generally increases iron absorption, but its effectiveness depends on the food type. During processing, minerals may be lost from foods (e.g., through leaching into the water) or added by the equipment used (e.g., iron from steel mixers or pipes). Bran and germ of grains have high amounts of minerals. However, the endosperm that remains after milling has low mineral content with a substantial

Table 6.2 Influence of food processing on minerals

Processing		Effects
Milling		Minerals lost from bran removal
Heat treatments	Blanching	Losses due to leaching into the water
	Pasteurisation	Few losses
	Sterilisation	Losses into brine/syrup
	Baking	Phytate destruction may increase absorption
	Frying	Iodine loses
Drying		Negligible influence
Freezing		Losses due to blanching before freezing losses in the drip after thawing
Fermentation		Reduction of phytate content
Extrusion		Bioavailability may increase or decrease
Packaging		Tin-cans may cause strong reducing conditions

reduction in Fe, Zn, Cu, Mg, and Sn. Nevertheless, the bioavailability of the remaining minerals may improve as phytic acid is also reduced. Steaming results in lower losses than boiling as minerals leach to the water, although steaming is not always possible. However, mineral bioavailability generally increases with cooking because of their improved solubility due to cell wall disruption, protein denaturation, and organic acids release that help with their solubilisation. Freezing and drying have a negligible influence on mineral content and absorption. Fermentation improves bioavailability due to lactic acid production, which enhances mineral solubility. Food storage has negligible effects on mineral content and bioavailability. Heat processing may influence bioavailability by changing mineral solubility and by destroying food constituents that modify their availability. Maillard reaction products may bind Zn, making it unavailable for absorption. Extrusion of cereals may increase Fe content due to contamination from the extruder. Additionally, an increase in mineral bioavailability because of phytate reduction is also possible, although the relationship is quite complex and depends on the food formulation. The following table presents a summary of the influence of various food processes on the mineral content of food (Table 6.2).

6.5   Learning Activities

6.5.1   Multiple-Choice Questions

1. The lipid-soluble vitamins are:

    (a) A, D, E and K
    (b) A, D, E and L
    (c) A, D, E and B
    (d) A, D, E and C

2. Vitamin A degrades with a similar mechanism as the:

   (a) Oxidative degradation of fatty acids
   (b) Hydrolysis of proteins
   (c) Enzymatic demethylesterification of pectin
   (d) Saturation of double bonds in fatty acids

3. Pantothenic acid has two parts in its structure that are:

   (a) Pantoic acid and β-alanine
   (b) Pentoic acid and β-alanine
   (c) Pantinoic acid and α-alanine
   (d) Pentoic acid and α-alanine

4. Which three structural parts are needed to form folate?

   (a) Pteridine, o-aminobenzoic acid and L-glutamic acid
   (b) Pteridine, p-aminobenzoic acid, and L-glutamic acid
   (c) Pteridine, m-aminobenzoic acid and L-glutamic acid
   (d) Pteridine, m-aminobenzoic acid and D-glutamic acid

5. Biotin in food is protein-bound and is called:

   (a) Biocysteine
   (b) Biocystine
   (c) Biocyteine
   (d) Biocytin

6. Ash, the inorganic residue from the incineration of food, contains:

   (a) All macrominerals
   (b) All trace-elements
   (c) All macrominerals and trace-elements
   (d) All macro minerals, trace elements and vitamins

7. Phytic acid has a very strong binding affinity for:

   (a) Ca, Fe and Zn
   (b) Ca, Cu and Zn
   (c) Fe, Na, and Cu
   (d) Cu, Zn and Mg

8. "The availability and reactivity of minerals depend on their water solubility". This statement is:

   (a) True
   (b) False

9. Ligands and minerals form complexes termed:

   (a) Chelites
   (b) Cheloates
   (c) Cheliceral
   (d) Chelates

10. In minerals, bioavailability is more important than content, as high content does not guarantee absorption. This statement is:

    (a) True
    (b) False

6.5.2   Short Answer Questions – Further Reading

1. **Online activity** Look online for the terms Daily Reference Intake (DRI), Recommended Dietary Allowance (RDA), and Daily Value (DV) and find the values for Vitamin A.

2. Write the structures of α-, β-, γ- and δ- tocopherols and tocotrienols and identify the differences between them.

3. List and discuss the causes of vitamin loses during food processing.

4. **Advanced, Online activity** Search online for the term "coordination number", "monodentate", "bidentate", and "polydentate".

5. List and discuss the causes of mineral losses during food processing.

6. **Further reading** Find the article "Schmid, A., & Walther, B. (2013). Natural Vitamin D Content in Animal Products. *Advances in Nutrition*, 4(4), 453–462."

    • What are the requirements of vitamin D in our diet?
    • What are the most important animal sources of vitamin D?
    • Discuss the bioavailability of vitamin D from foods and supplements.
    • Discuss the influence of processing on vitamin D stability.

7. **Further reading** Find the article Hunt, J. R. (2003). Bioavailability of iron, zinc, and other trace minerals from vegetarian diets. *The American Journal of Clinical Nutrition*, 78(3), 633S–639S" and discuss the iron and zinc bioavailability in vegetarian diets.

6.5.3   Fill the Gaps

1. Vitamin A degradation may be accelerated in _____ foods where oxidation generally proceeds faster.

2. Dietary sources of _____ are insignificant compared to the amount prepared *via* the biosynthetic pathway.

3. Regardless of the processing, ~20% of _____ is generally lost.

4. Folic acid is very stable in _____ and _____ environments.

5. _____ is the addition of specific amounts of vitamins lost during processing.

6. The availability and reactivity of minerals depend on their water _____.

7. Ligands and metals form complexes termed _____,

8. _____ acid has strong binding affinity to the dietary minerals.

9. Phytochemicals like _____ and _____ also influence binding of minerals.

10. In minerals, _____ is more important than content.

Chapter 7
Colour Chemistry

Learning Objectives

After studying this chapter, you will be able to:

- Describe the colour of myoglobin with changes in iron oxidation
- Describe colour changes of chlorophyll during processing
- Describe carotenoid formation and colour changes during processing
- Describe colour changes of anthocyanins with pH
- Describe the chemical structures betalains
- Describe the structure of azo dyes

7.1 Introduction

Taste, flavour, texture, and colour are foremost for food quality and sensory accept-ability, with freshness, taste, and colour being the top three criteria that define food quality in consumer studies. Colour may indicate chemical changes of food, flavour, or texture (e.g., ripening stage or crust colour of bakery products) and is the first assessment of satisfaction even before consuming the product. Control of colour intensity and understanding colourant interactions with other food components during processing and storage is essential to ensure high food quality and prolonged shelf life. Adding colourants is done due to variations in the colour of food ingredients (e.g., seasonal variations), changes occurring during processing (e.g., due to heat degradation), or to compensate for colour losses due to exposure to light, oxygen, or moisture during storage. In addition, colours that occur naturally (e.g., red in a strawberry-containing drink) may need to be enhanced or provide colour to colourless foods and create a unique sensory experience (e.g., green in mint fla-voured ice cream).

© The Author(s), under exclusive license to Springer Nature Switzerland AG 2021
V. Kontogiorgos, *Introduction to Food Chemistry*,
https://doi.org/10.1007/978-3-030-85642-7_7

Colour is the sensory quality of an object perceived with the eyes from the light reflected by the object. Colour perception results from the stimulation of *photoreceptor cells* (*cones* and *rods*) by electromagnetic radiation with wavelengths between 380 and 780 nm. Radiation within this range is called the *visible region of electromagnetic radiation*. Radiations outside this range (e.g., ultraviolet or microwave) are not detectable by the human eye. Rods are sensitive to lightness and darkness and cones to red, green, and blue, with the human eye being most sensitive at ~555 nm (green). The visible region of the electromagnetic spectrum is the *stimulus* for colour perception. However, a combination of the light source, object properties, the anatomy of the eye and the brain results in a unique and individual sensory experience. It is estimated that the human eye can distinguish about ten million different colours. Light source (e.g., sunlight, fluorescent, tungsten, neon, or blacklight) is an important parameter that dramatically influences colour perception. For example, rice that appears white under sunlight (consisting of ultraviolet, visible, and infrared radiations) appears purple when it is illuminated with blacklight (consisting of long-wave ultraviolet light (UV-A), very little visible light and no infrared radiation). Background differences also play an important role in colour perception, as colours may appear different depending on their surrounding colours and shapes (e.g., bright *vs* dark background or the Chubb illusion). Other parameters, such as sample size (the same colour looks different in objects of varying size), angle of observation (colours may be different from different angles of observation), or observer differences (i.e., the sensitivity of an individual's eyes and brain) may also influence colour perception.

7.2   Interaction of Light with Food

Objects are characterised by the amount of light they *emit, reflect,* or *transmit* at each wavelength of interest. When light is incident on an object, a part of it is absorbed (i.e., "stays inside the object"), reflected (i.e., bounces back to the observer) or transmitted (i.e., goes through the object) (Fig. 7.1). Sometimes, a chemical

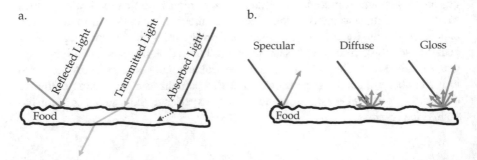

Fig. 7.1 (**a**) Modes of light interaction with the food surface, and (**b**) different types of reflection

found in food may emit light, as in *fluorescence dyes*. For example, when tonic water is illuminated with blacklight, it emits bright blue light because of quinine found in the formulation. Reflected light may present different forms of reflectance depending on the surface properties of food. In *specular reflection* (i.e., mirror-like), the angle of incidence on the surface is equal to the angle of reflection. Ideal specular reflection is never observed in foods. In *diffuse reflection,* the light beam is scattered at many angles rather than at just only one. A combination of both modes of reflection (specular and diffuse) results in *gloss reflectance* (Fig. 7.1).

In practical terms, a food that strongly scatters light in all directions appears matte (i.e., not shiny), whereas when specular reflection is dominant, it appears glossy (i.e., shiny). For instance, the external cell layer of fruits (i.e., cuticle) consists of *cutin,* a waxy material imparting fruits with a strong glossy appearance (e.g., apples, cherries, etc.). Degradation of the cutin layer during storage because of biochemical changes or abrasion during transportation may result in loss of gloss. *Cloudiness* and *translucency* are two terms that are related to transmitted light. Cloudy liquids contain particles or droplets that scatter light in all directions (e.g., cloudy lemonade, milk, or cloudy apple juice). Transparent liquids do not contain particles, or particles are very small in the range of nanometres being unable to scatter light. As a result, the fluid appears clear (e.g., clear apple juice or cranberry juice).

Colour Measurement

A suitable mixture of red, green, and blue lights can match any coloured light. This observation forms the basis of colour measurement (i.e., colourimetry) and *colour classification systems* (or *colour spaces*). There are various colour spaces available depending on the underlying fundamental principles used to describe them and the desired application (e.g., CIE 1931 XYZ, Munsell system, Adobe RGB and many others). CIELAB colour space, abbreviated as L*a*b*, is a system of colour classification that approximates human vision and is most frequently used in both industry and research for colour measurements. Colour is expressed as three values: L* for the lightness that ranges from 0 (black) to 100 (white), a* from that ranges from green (−128) to red (+127), and b* that ranges from blue (−128) to yellow (+127). L*a*b* values can be related to pigment concentration, and relationships may be drawn between concentration *vs* intensity of the colour.

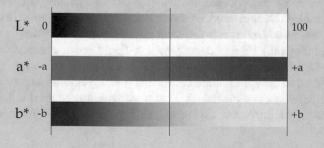

In L*a*b* system, numerical changes in these values correspond to approximately the same amount of visually perceived differences. L*a*b* values can be converted to a single colour function called *colour difference* (ΔE*) using the following equation:

$$\Delta E^* = \sqrt{\Delta L^{*2} + \Delta a^{*2} + \Delta b^{*2}}$$

with $\Delta L^* = L^*_{sample} - L^*_{standard}$, $\Delta a^* = a^*_{sample} - a^*_{standard}$, and $\Delta b^* = b^*_{sample} - b^*_{standard}$. Other equations are also available that take more parameters into account, but nevertheless, the one presented above provides useful information about the differences between two colours. For example, quite frequently, the difference in colour during processing (e.g., frying or baking) needs to be closely monitored and controlled. Two colours are different when $\Delta E^* \geq 1$. When $\Delta E^* > 3$, this is perceived as a significant colour deviation. However, depending on the food, occasional batch-to-batch colour variations may be tolerated with $\Delta E^* > 4$ (e.g., baked, or fried foods) in contrast to others that slight deviations may signify product rejection (e.g., meat products). Colourimetry is used extensively in the food industry in quality control to ensure consistent colour, product development (e.g., new processes or ingredient replacement), and shelf life determination to monitor colour changes during storage, and in consumer studies to identify which colour is preferred.

7.3   Colour Chemistry

Food colour results from the presence of *colourants* divided into *dyes, lakes, and pigments*. Colourant is a generic term that describes any substance that colours something. A dye is a colourant that is typically organic water-soluble and can molecularly dissolve to form a coloured solution. Dyes are converted into dispersions called *lakes,* usually using aluminium hydroxide. The dye is bound to aluminium hydroxide particles, making it water-insoluble. However, it can be dispersed in fats, oils, and other non-aqueous formulations (e.g., cake mixes). For example, a blue sports drink is coloured with a dye, whereas a blue chewing gum with a lake. *Pigments* are inorganic compounds, usually oxides that are not water-soluble. Examples include titanium dioxide used in dairy or iron oxides used in cake mixes. Colourants can be converted into different physical forms depending on food composition and they may be applied as powders, liquid concentrates, dispersions, or emulsions. Despite the above definitions, in life sciences the word "pigment" is commonly used to describe natural colourants (i.e., myoglobin, chlorophyll, carotenoids, anthocyanins, and betalains). This word will also be used in the following sections, but it is important to know and appreciate the differences between the terms.

Colourants used in food fall into two classes: *natural* and *artificial,* both with their advantages and disadvantages. There are *five classes of natural pigments* and are classified most frequently based on their aqueous or lipid solubility or based on their chemical structure. *Anthocyanins* and *betalains* are water-soluble, whereas *carotenoids* and *chlorophylls* are lipid-soluble and are of plant origin. However, some carotenoids may also be found in non-plant foods. For example, the colour of crabs or shrimps is due to the presence of various carotenoids, most frequently astaxanthin and β-carotene, whereas the colour of egg yolk is due to carotenoids lutein and zeaxanthin. *Myoglobin* is the only natural pigment that is derived from animal sources and is water-soluble. In terms of chemical structure, natural pigments are classified into four major groups: *isoprenoid derivatives* (carotenoids and xanthophylls), *tetrapyrrole derivatives* (chlorophylls, heme), *benzopyran derivatives* (anthocyanins, flavonoids, tannins) and *betalamic acid derivatives* (betalains).

7.3.1   Myoglobin

Myoglobin (Mb) is a protein found in muscle cells and is responsible for meat colour. This protein has a prosthetic group called haem (or *heme*) containing iron and can react with oxygen (Fig. 7.2).

In contrast to haemoglobin found in red blood cells and essentially in the entire body, Mb is only found in muscle tissues of animals. It should be noted that since blood is drained during slaughtering, the colour of meat is *almost exclusively due to the presence of Mb* and any residual haemoglobin plays only a minor role. The *state of iron oxidation* in haem (i.e., ferrous (Fe^{2+}) or ferric (Fe^{3+})) determines the colour of meat. The colour of Mb in fresh meat and under the complete absence of oxygen is purple and is termed *deoxy-Mb* (Fe^{2+}) (Fig. 7.3). When Mb interacts reversibly with oxygen but without changes in the state of Fe oxidation, it forms oxymyoglobin (*oxy-Mb*, Fe^{2+}) that has a bright red colour, the typical appealing colour of red meat. When iron is oxidised to Fe^{3+}, metmyoglobin (*met-Mb*, Fe^{3+}) is formed that

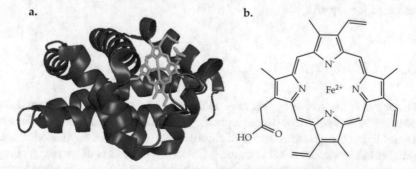

Fig. 7.2 (**a**) Structure of myoglobin. In green is haem located within the protein, and (**b**) structure of haem. Structure of myoglobin was drawn in PyMOL based on PDB accession code 1WLA

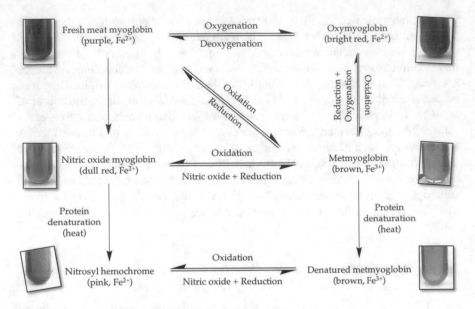

Fig. 7.3 Colour of myoglobin with changes in iron oxidation

has a brown colour. Reduction of Fe^{3+} to Fe^{2+} reverts colour to red, but this process is not used in practice. The addition of nitrates and nitrites used in cured meat products creates nitrous oxide (NO) that interacts with Mb yielding *NO-Mb* (nitroso myoglobin, Fe^{2+}) with pink colour after application of heat (e.g., cooked hams). Carbon monoxide (CO) may also be applied to improve colour that proceeds with the formation of *CO-Mb* (Fe^{2+}) (carboxy myoglobin, bright red).

Green colours may also be formed because of the presence of sulfmyoglobin (H_2S), choleglobin, oxidised porphyrins, or excess of NO ("nitrite burn"). Colour intensity depends on Mb concentration and it varies between different species in the following the order: chicken (0.02 mg/g) < pork (2 mg/g) < lamb (6 mg/g) < beef (8 mg/g). Additionally, older animals have a higher concentration of myoglobin than young, while muscles used for motion have a higher concentration of Mb than those used for support.

7.3.2 Chlorophyll

Chlorophyll is the green pigment in plant tissues (e.g., leafy vegetables) and allows plants to absorb light energy through photosynthesis. Minor differences in the structure and light absorption properties of chlorophylls (e.g., chlorophyll *a*, chlorophyll *b*, etc.) result in various shades of green. Chlorophyll consists of two main components: a porphyrin ring and a phytol *tail*. Porphyrin ring is similar to haem but with bound Mg^{2+} instead of Fe^{2+} (Fig. 7.4).

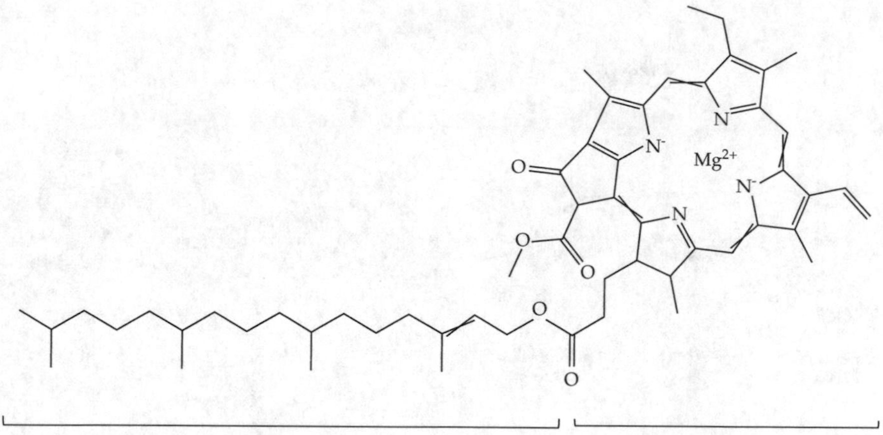

Phytol tail   **Porphyrin ring**

Fig. 7.4 Structure of chlorophyll *a*

Phytol tail is the hydrophobic part of the molecule and gives chlorophyll its lipophilic characteristics. Any type of food processing causes some changes to chlorophyll pigments. Food composition and processing conditions, e.g., NaCl, pH, temperature, and heating duration, all influence colour. Exposure to severe heat and/or acidic conditions during canning results in loss of Mg^{2+} generating the olive-brown pheophytin and *pyropheophytin* pigments. *Chlorophyllase* activity, induced by blanching, results in the formation of water-soluble chlorophyllides (dark green) that further degrade through thermal processing and/or acidification to pheophorbide pigments (olive green) (Fig. 7.5).

7.3.3 Carotenoids

Carotenoids are fat-soluble compounds having yellow, orange and red shades and are the most widespread pigments. There are two carotenoids classes, *hydrocarbon* carotenes and *oxygenated* xanthophylls, with over 600 different compounds in these two classes. Carotenoids are formed with head-to-tail polymerisation of isoprene units that results in the presence of conjugated double bonds (Fig. 7.6). Head to tail polymerisation proceeds with the linkage of C-1 of the first isoprene to the C-4 of the second isoprene, also known as *1-4 linked terpenes*. Potentially the links could also be 1-1 (tail-tail) and 4-4 (head-head). However, in most naturally occurring terpenes, the isoprene rule states that there are no 1-1 or 4-4 links. A terpene that does not obey the isoprene rule is called *irregular terpene*, and the most common example is *β*-carotene (Fig. 7.7).

The absorption of carotenoids shifts to longer wavelengths as the number of conjugated bonds increases (i.e., it becomes redder). Carotenoids require at least

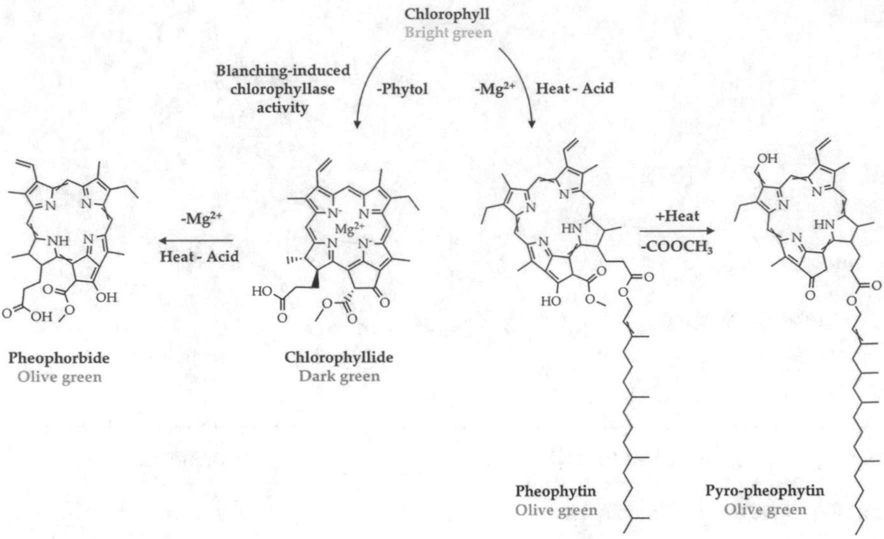

Fig. 7.5 Colour changes in chlorophyll with processing conditions

Fig. 7.6 Head to tail polymerisation of isoprene to form myrcene. Myrcene is the simplest mono-terpene and a main component the essential oils

nine conjugated double bonds before the appearance of yellow colour. The configuration of the double bond in carotenoids is *trans* type, although *cis* may also be present. Carotenoids with *trans* bonds have deep colour, whereas the increase of *cis* configuration results in colour fading. This configurational change may occur during processing and storage under the influence of light, oxygen, acid, and heat. Common unit operations during food processing do not particularly influence the colour of carotenoids, although this depends on the specific food formulation. Blanching destroys enzymes that bleach carotenoids, thus preserving their colour in frozen foods. More severe heat treatments (e.g., sterilisation) do not generally influence the carotenoid colour. In contrast, dehydration degrades pigments, and dry fruits and vegetables usually have faded colour compared to their fresh counterparts.

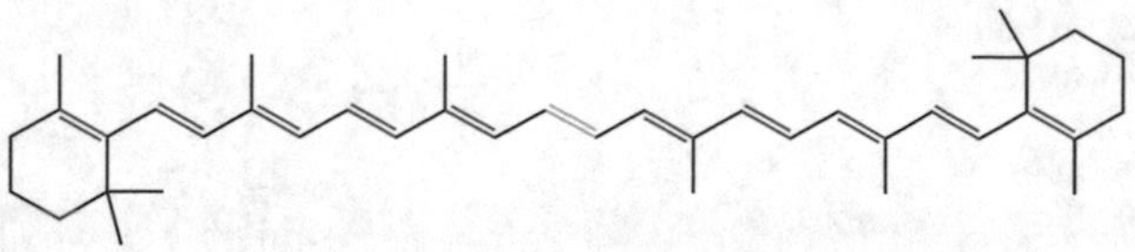

β-Carotene

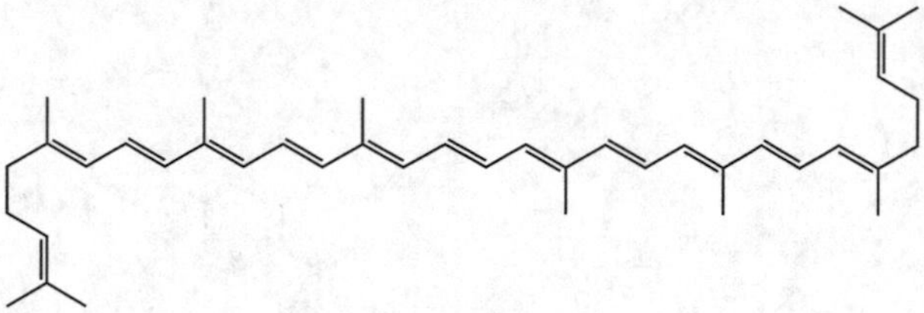

Lycopene

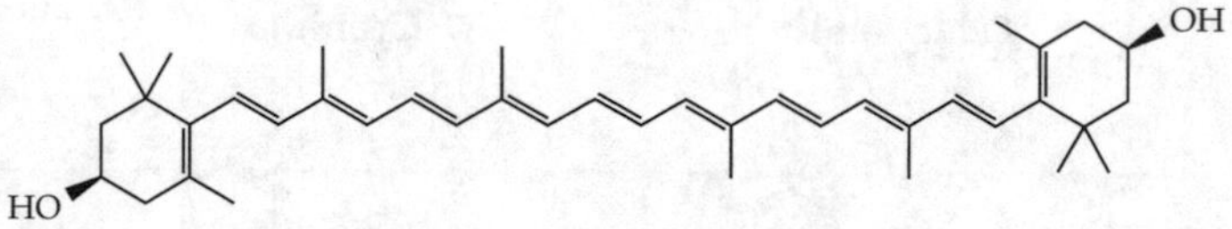

Zeaxanthin

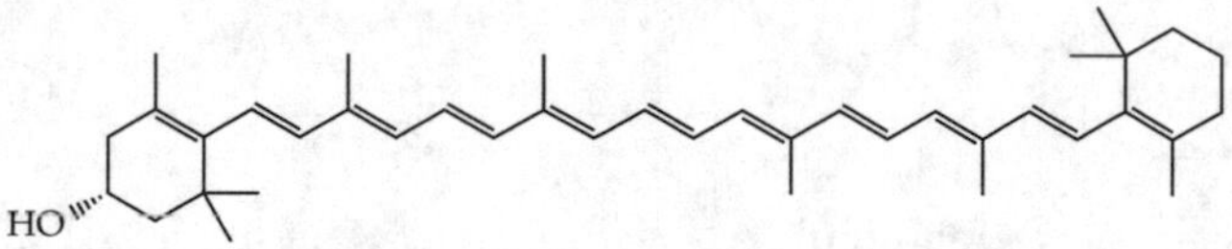

Cryptoxanthin

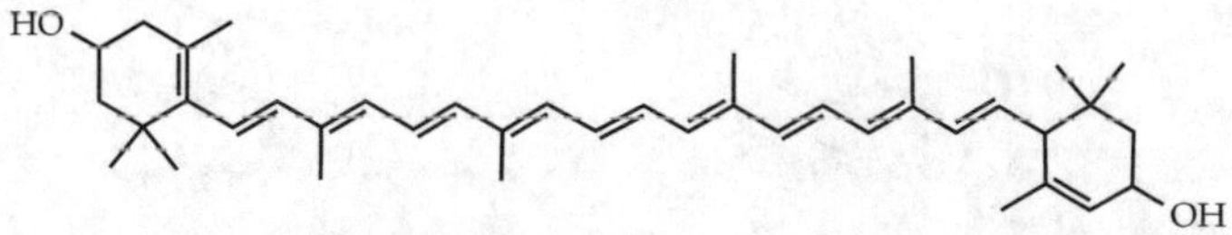

Lutein

Fig. 7.7 Examples of carotenoid structures. *β*-carotene and lycopene are hydrocarbon carotenes, whereas the rest are oxygenated xanthophylls. The conjugated bonds of *β*-carotene are shown in red. The irregular double bond formed by 1-1 polymerisation (tail-tail) is shown in green

7.3.4 Anthocyanins

Anthocyanins are found in flowers, fruits, and vegetables, with more than 700 different molecules in this group. The colour varies between blue, pink, or violet shades and has *strong pH dependency* due to the molecule being amphoteric. The building block of anthocyanins is known as flavylium cation (Fig. 7.8), and they usually contain one molecule of glucose or galactose.

Without the monosaccharide, the molecules are referred to as anthocyanidins. Flavylium has several hydroxy and methoxy substituents, and the number of –OH substitution of the R- groups on flavylium cation results in different colours.

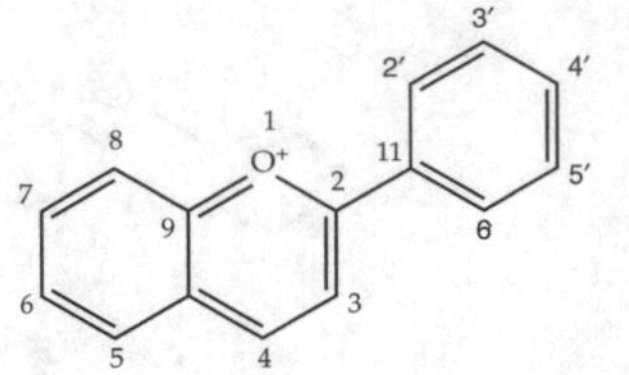

Fig. 7.8 Flavylium cation
is the building block of
anthocyanins. The
numbering of the atoms is
also shown

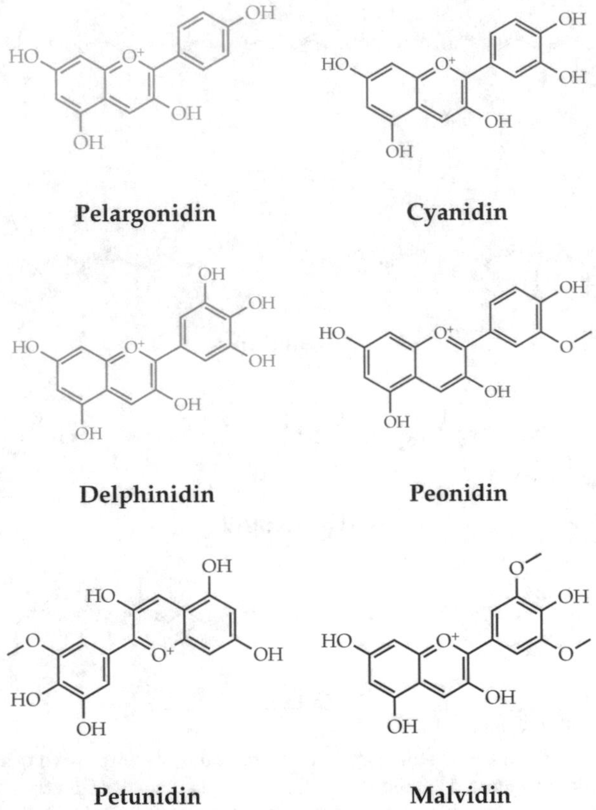

Pelargonidin

Cyanidin

Delphinidin

Peonidin

Petunidin

Malvidin

Fig. 7.9 Common anthocyanins occurring in food. The colour of the structure indicates its approximate colour. The colour is highly sensitive to pH

More –OH group substitutions result in blue shift, whereas the addition of methoxyl groups results in red shift. The six anthocyanidins *pelargonidin*, *cyanidin*, *delphinidin*, *peonidin*, *malvidin*, and *petunidin* are the most commonly occurring in foods (Fig. 7.9).

Hydroxyl groups generally decrease the stability of the pigment. In contrast, a higher number of -OMe substituent groups in the structure increases its stability. At pH < 2, the orange or red flavylium cation predominates. As the pH increases, two competing reactions may occur between the hydration reaction on position 2 of the flavylium cation and the proton transfer reactions of the acidic hydroxyl groups. The first reaction gives colourless *carbinol pseudobases* (pH ~5), which can undergo ring-opening to pale yellow or colourless chalcones (pH ~6). The second reaction gives rise to violet *quinoidal bases* (pH ~4). Further deprotonation of the quinoidal base can occur at pH between 6 and 7 to form bluish quinonoid anions (Fig. 7.10). **Anthocyanins are present as a mixture of equilibrium forms**. As pH increases above ~2.5, the most predominant form at any given pH in the equilibrium mixture follows the order (more) carbinol > chalcone > quinoidal bases (less).

Anthocyanins are very unstable and sensitive to many factors acting in coordination depending on the food. Strategies to retain anthocyanin colour need to take into consideration every aspect of processing and storage. Generally, heat processing shifts the equilibrium towards chalcones with oxidation and cleavage of covalent bonds by heat resulting in unpleasant yellow or brown colours. Oxygen degrades colour from purple to brown and must be removed from packaging (e.g., vacuum packaging). Light also accelerates degradation, and packaging of food rich in anthocyanins should be in containers impermeable to light. High concentration of sugars (e.g., jams) stabilise anthocyanins because of a_w lowering. However, a low concentration of sugars accelerates degradation. Anthocyanins degrade faster in the presence of ascorbic acid, which is important for food formulations made with fruits. Flavonoids are water-soluble compounds with primarily yellow colour and chemical structures similar to anthocyanins, with *quercetin* being the most widely distributed (Fig. 7.11). They occur widely in plant materials, with over 6000 compounds known to date. They are generally more stable than anthocyanins. They are not used as industrial ingredients, but the yellow colour of onions, cauliflower and potato is attributed to flavonoids.

7.3.5 Betalains

Betalains are red and yellow pigments divided into betacyanins (red, violet) and betaxanthins (yellow, orange). Interestingly, they only occur in the Caryophyllales order of flowering plants such as beets, amaranth and various cacti. Betacyanins have a similar colour to anthocyanins but *are not affected by pH* whereas betaxanthins are less common. Betalains are nitrogen-containing compounds formed by condensation of betalamic acid with an amine. Substitution with different amines results in distinct red shades (Fig. 7.12). In addition, monosaccharides may also be present, affecting the colour in different ways. For example, *betanin* from red beets contains glucose. Betalains are stable at pH 3.0–7.0, but heat processing in acidic conditions degrades them resulting in yellow colours. Oxygen and light result in bleaching and packing should be carried out under vacuum. Some naturally found enzymes (e.g., peroxidases) may catalyse oxidative degradation of betalains.

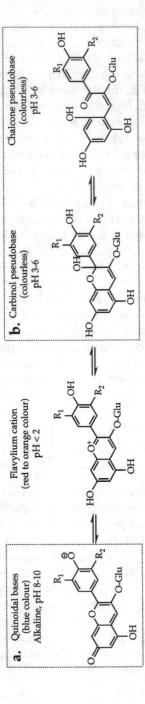

Fig. 7.10 Colour changes of anthocyanins with changes in pH. The starting point is the flavylium cation in the middle at pH < 2. (**a**) Violet quinoidal bases start forming at pH ~4. Further increase of pH between 6 and 8 results in blue colour, and (**b**) colourless carbinol pseudobases form at pH ~5 which can change to pale yellow or colourless chalcones at pH ~6. Glu is glucose

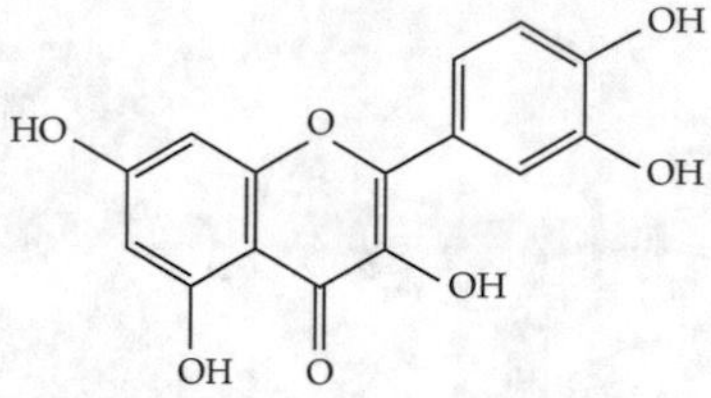

Fig. 7.11 Structure of quercetin

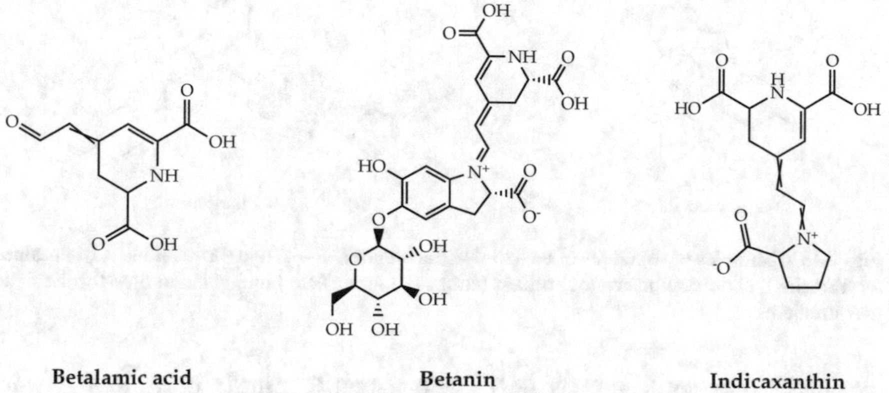

Betalamic acid   Betanin   Indicaxanthin

Fig. 7.12 Structure of two betalains. Betalains are derivatives of betalamic acid

Finally, other natural pigments may be obtained from various natural sources frequently without purification e.g., *spirulina extracts* (phycocyanins), *saffron* (crocin, carotenoid), *annatto* (bixin and norbixin, carotenoids), *cochineal* (carminic acid) or *turmeric* (curcumin).

7.3.6 Artificial Dyes

Artificial dyes are needed because the range of shades and quantities from natural sources are limited. In addition, synthetic dyes are made to provide a range of alternatives for food manufacturers to improve visual attractiveness. In addition, chemically synthesised dyes have many advantages over natural colourants such as a brighter and broader range of shades, greater stability during processing and storage (e.g., against heat, light, oxygen etc.) and are generally cheaper. Caramel colours that are produced through caramelisation reactions have been described in Sect. 5.3.1.2.

The major group of artificial dyes are the azo-dyes, followed by *triaryl methane* and *xanthine* dyes, all entirely synthetic. Azo dyes are derivatives of *diazene* (H–N=N–H). The N=N group found in their structure is called *azo group*, with some dyes having more than one azo group (i.e., di-azo dyes). They have a general

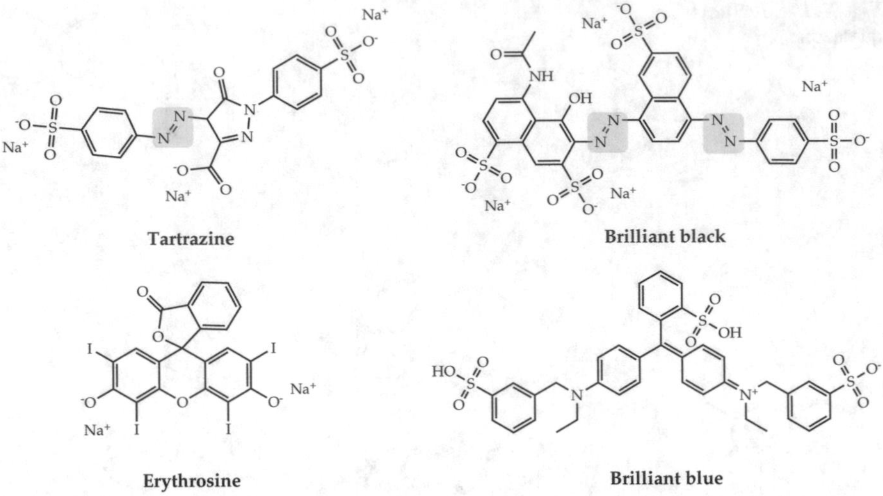

Tartrazine   Brilliant black

Erythrosine   Brilliant blue

Fig. 7.13 Examples of structures of an azo-dye (tartrazine), di-azo dye (brilliant black), xanthine (erythrosine), and triaryl methane (brilliant blue). Azo groups are highlighted in blue for the first two structures

structure written as $R - N{=}N - R'$, with R and R' usually being *aryl groups* (Fig. 7.13). The list of authorised artificial dyes undergoes safety assessment from relevant authorities depending on the regulations of each region. Depending on the region, around 10 to 15 artificial dyes are currently authorised for use in foods. The complete lists are available online on the websites of the relevant institutions (e.g., EFSA, FDA etc.). The dyes in the lists are periodically re-evaluated and may change when new toxicological data on their safety become known.

7.4   Learning Activities

7.4.1   Multiple-Choice Questions

1) Specular and diffuse reflectance of light on fruits surface results in:

   a) Gloss appearance
   b) Matte appearance
   c) Fluorescent appearance
   d) Translucent appearance

2) In the L*a*b* system of colour, classification b* is:

   a) Colour variation from blue to yellow
   b) Colour variation from red to green
   c) Colour variation from white to black
   d) Colour variation from blue to green

3) Oxygenation of myoglobin results in:

   a) Oxymyoglobin
   b) Metmyoglobin
   c) Denatured myoglobin
   d) Denatured metmyoglobin

4) Oxidation of myoglobin results in:

   a) Oxymyoglobin
   b) Metmyoglobin
   c) Denatured myoglobin
   d) Denatured metmyoglobin

5) Chlorophylls are water-soluble components that are sensitive to pH. This statement is:

   a) True
   b) False

6) The olive-green colour of green vegetable after blanching is due to the loss of:

   a) Mg
   b) Fe
   c) Ca
   d) Zn

7) Anthocyanins have a very similar structure to carotenoids, but they give blue/purple shades. This statement is:

   a) True
   b) False

8) Increasing the pH of a delphinidin solution changes its colour from:

   a) Blue to green
   b) Blue to red
   c) Green to blue
   d) Red to blue

9) Betalains are typical derivatives of the flavylium cation. This statement is:

   a) True
   b) False

10) Porphyrin ring is present in:

    a) Chlorophylls and heme
    b) Flavylium cations and heme
    c) Betalamic acid and chlorophyll
    d) β-carotene and betacyanins

7.4.2   Short Answer Questions – Further Reading

1) Discuss the purpose of colouring food and the parameters that affect colour perception.

2) Explain why some meat products (e.g., cooked ham, sausages) have pink colour.

3) Discuss the reactions that occur in chlorophyll during blanching and heating in an acidic environment.

4) **Online activity:** Search online for the biosynthetic pathway of carotenoid formation in plants and the importance of phytoene.

5) **Online activity:** Search for the term "apocarotenoid", "bixin", and "norbixin". In which natural food colouring we find these compounds?

6) Draw and discuss the reactions that occur in anthocyanins with changes in pH.

7) **Online activity:** Find the colourant with "E numbers" E160d, E161b and E162.

8) **Online activity:** Find the structures of Ponceau S, Allura Red, and Brown HT and identify the azo group.

9) Further reading: Find the article "Pathare, P. B., Opara, U. L., & Al-Said, F. A.-J. (2013). Colour Measurement and Analysis in Fresh and Processed Foods: A Review. *Food and Bioprocess Technology*, 6(1), 36-60."

   • Define the following terms "Total Colour Difference", "Whiteness Index", and "Yellowness Index".
   • How can we quantify colour changes in browning?
   • What types of empirical mathematical modelling are available to follow colour degradation kinetics?

7.4.3   Fill the Gaps

1) A fluorescent dye found in food is _____.

2) The glossy appearance of some fruits is because of _____.

3) The state of iron _____ in _____ determines the colour of meat.

4) The colour of metmyoglobin is _____.

5) The colour of nitroso myoglobin is _____.

6) Exposure of chlorophyll to heat and acidic conditions results in loss of _____ generating the olive brown _____ and _____ pigments.

7) Carotenoids are formed with polymerisation of _____ units that result in _____ double bonds.

8) Carotenoids with ___ bonds have deep colour, whereas increase of ___ configuration results in colour fading.

9) Anthocyanins without the monosaccharide are referred to as _____.

10) The colour of quinoidal bases is _____.

11) _____ have similar colour to anthocyanins but are not affected by ____.

12) The N=N group is found in the structure of _____ dyes.

Chapter 8
Flavour Chemistry

Learning Objectives

After studying this chapter, you will be able to:

- Recognise the chemical structures of major compounds responsible for flavour
- Describe the chemical structure of compounds with sweet and bitter tastes
- Discuss how flavour is generated and its sources
- Discuss how flavour is delivered and its stability in foods

8.1 Introduction

Flavour is the sensory impression of food determined mainly by the *chemical senses* of *taste* and *smell*. They are called chemical senses because molecules present in the food trigger them when they associate with the tongue's receptors. The *trigeminal senses* detect chemical irritants, temperature changes, and texture, which are also critical to the overall flavour perception. Chemicals that prompt a chemical sense are naturally found in the raw materials, but also, they are frequently added to food. There are three main reasons to add flavourings: (i) to add a flavour that exists but is not typical of the product, for example, a flavoured mineral water with citrus extracts, (ii) to add a flavour that has been lost or modified during processing, e.g., flavours lost from a fruit drink during heat treatments, and (iii) to create a flavour that does not exist, e.g., in cocoa or coffee bean roasting.

Flavours can be either *natural*, *process*, or *artificial*. The exact definitions vary depending on the legal framework of each region (e.g., FDA, EU, etc.). Regardless of the legal definitions, natural flavours are extracted from plant or animal materials and are not chemically modified any further, whereas artificial flavours are obtained by chemical synthesis. The differences between the two can be illustrated with vanilla flavour. The natural vanilla flavour is extracted from vanilla beans and

© The Author(s), under exclusive license to Springer Nature Switzerland AG 2021
V. Kontogiorgos, *Introduction to Food Chemistry*,
https://doi.org/10.1007/978-3-030-85642-7_8

diluted in ethanol. The five main compounds contributing to vanilla flavour are vanillin, 4-hydroxy benzaldehyde, 4-hydroxy benzoic acid, vanillic acid, and 3-methoxy benzaldehyde. In addition to these five components, the extract has over 150 compounds that contribute to natural vanilla flavour. However, we may chemically synthesise only these five main molecules and mix them in ethanol in a certain proportion. The result is a synthetic vanilla flavour (artificial flavour) that smells like the natural but only contains these five compounds. Most commercial flavours are chemically synthesised and formulated to resemble their natural counterparts rather than extracted from natural sources. *Process flavours* are produced during processing (e.g., caramelisation) or made by some processing technique, usually thermal or enzymatic (e.g., protein hydrolysates). see Sect. 8.3).

Generally, flavours have four characteristics that make them unique and distinguishable: the *primary* and *secondary* characters, the *complexity*, and the *balance*. The *primary character* is essential to recognising target food and constitutes the basic component of the flavour. It is impossible to create a realistic flavour without some contribution from the primary character. The *secondary character* is not essential for recognition but contributes important descriptive characteristics. For example, pure benzaldehyde rarely invokes the flavour of cherries, even though it is a major chemical of cherry flavour. However, a cherry extract is easily distinguishable, as it contains several secondary compounds making the extract flavour characteristic of cherries. The third quality of flavours is their *complexity* that varies radically depending on the purpose of use. Essentially, complexity refers to the number of different chemicals that simultaneously coexist in the flavour preparation and contribute to the recognition and quality of the flavour. The level of complexity may vary from perhaps as few as 15 components in simple fruit flavours to up to 300 in the most complex flavours of processed food (e.g., Maillard reactions). Finally, *flavour balance* is an important characteristic, as synthetic flavours may become unbalanced when a compound overwhelms the others. It is generally challenging to mix artificial flavours to replicate a natural counterpart, as it may easily result in unbalanced flavours. Natural flavours are usually balanced.

8.2 Flavour Chemistry

The five basic tastes (i.e., sweet, salty, sour, bitter, and umami) are *chemical senses*. In this case, a chemical substance (e.g., quinine) triggers a unique neural response by the sensory receptors (taste buds), resulting in taste perception (i.e., bitterness) that finally translates to a behavioural response (in the example of bitterness, it is usually repulsion). Chemical compounds interact with *taste-receptors* and transduce signals to the brain. Frequently, different chemicals result in similar tastes. Consequently, there must be a unique *structure-function relationship* between the structure of the chemical stimulus and the receptors of the taste buds.

The most well studied structure-function relationship is that of *sweet molecules* that is generally well understood and has allowed the creation of many artificial non-calorific sweeteners with structures quite different than those of sweet carbohy-drates (e.g., glucose). Initially, a sweet-tasting molecule (also termed *glycophore*) *must be water-soluble*, as this allows strong association through hydrogen interactions between the molecule and the taste receptor. In addition, a sweet molecule must have the correct molecular *shape* and the proper *electronic distribution* that allows it to interact with the taste receptor. The current theory of sweetness, known as *AH-B-γ theory* or the *triangle theory*, suggests that the spatial arrangement of atoms on interaction with tongue receptors is triangular (Fig. 8.1).

The sweet molecule binds to the taste receptor site in a complementary manner through hydrogen bonding and hydrophobic interactions. The AH^+ area contains functional groups such as hydroxyl, amine, or some other groups with hydrogens available to hydrogen-bond to some partially negative atom such as oxygen. The B^- area contains functional groups such as hydroxyl, carboxyl, or some other group with partially negative oxygen available to hydrogen-bond to a partially positive atom, such as hydrogen, on the sweet molecule. Finally, the γ area is perpendicular to the other two regions interacting through hydrophobic interactions with the tongue receptors. Functional groups such as benzene rings and multiple $-CH_2$ and $-CH_3$ groups are usually found in the γ-area (Fig. 8.2). Typically, calorific sweeteners are carbohydrates, most frequently sucrose, glucose, fructose, and their mixtures. Sugar alditols also have a sweet taste but do not provide calories (e.g., sorbitol or xylitol). Other structures of sweet-tasting molecules include sulfur-containing structures such as *cyclamates* and *saccharin*, glycosides (*steviol glycosides*), peptides (*aspartame*) or even proteins (*thaumatin*).

Bitter *molecules* are closely related to sweet molecules from a molecular structure-receptor relationship. Sweet molecules need two polar groups and a non-polar for functionality, whereas bitter molecules *need one polar and one non-polar* to cause a bitter taste. Bitter molecules may also taste sweet depending on the stereochemistry of the compound resulting in a bitter-sweet taste. Some synthetic sweeteners have a *bitter aftertaste* that needs to be masked, with *bitter taste-masking technologies* being a current field of flavour research. Bitter compounds have a

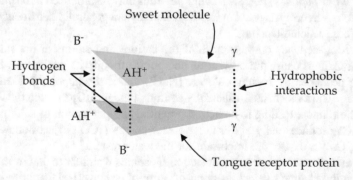

Fig. 8.1 The spatial arrangement of the appropriate groups of a sweet molecule should be triangular to interact with the sweet-taste receptors of the tongue. Yellow triangle: sweet molecule, grey triangle: tongue receptor

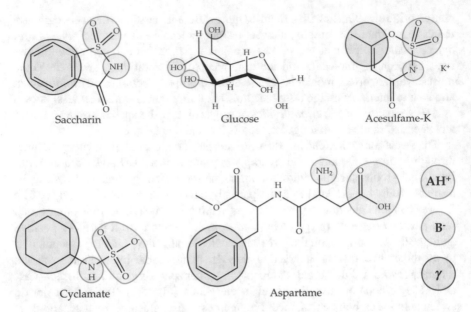

Fig. 8.2 Structures of some common sweeteners. Highlighted areas are the functional groups that interact with tongue receptors

lower detection threshold and are *less soluble in water than sweet molecules.*
Stereochemistry strongly influences bitterness, as in L-tryptophan that is bitter,
whereas D-tryptophan tastes sweet (Fig. 8.3).

Bitter compounds are structurally diverse but quite frequently are alkaloids.
Alkaloids are organic nitrogen-containing bases and can be classified into various
groups depending on the fine details of their structure. *Quinine*, frequently used in
beverages (e.g., tonic water), is a *quinidine alkaloid*, whereas *caffeine* incoffee and
theobromine incacao are *xanthine alkaloids*. Other bitter structures include *glycosides* (e.g., naringin in grapefruit), *phloroglucinol derivatives* (e.g., humulone in
beer), or *bitter peptides* (Fig. 8.4). Bitter peptides may be produced in fermentations
due to protein hydrolysis (e.g., in cheese) and they are generally detrimental for the
product (e.g., leucine-leucine).

Salty *taste* depends on the nature of the cations and anions of the salt crystal
structure. Cations cause the salty taste, and anions modify its perception. Saltiness
is a taste produced by sodium chloride (NaCl) that is the major salt used in food
processing. The ions of NaCl, especially sodium, pass directly through the ion channels of the tongue, leading to an action potential and perception of saltiness in the
brain. KCl produces salty and bitter tastes whereas HCO_3^- (bicarbonate), SO_4^{2-}
(sulfate), Ca^{2+} and Mg^{2+} are important for the water taste.

Sour *taste* detects acids with a detection mechanism similar to that of salty taste.
Hydrogen ion channels detect the concentration of protons (H^+) that have been dis-
sociated from an acid. There is no strong correlation between pH and sour taste.
However, different organic acids have different tastes. Citric acid provides a "fresh

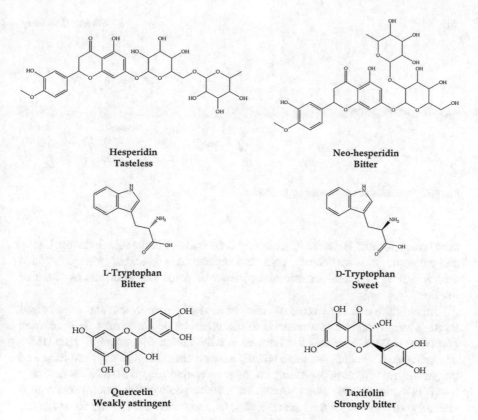

Hesperidin
Tasteless

Neo-hesperidin
Bitter

L-Tryptophan
Bitter

D-Tryptophan
Sweet

Quercetin
Weakly astringent

Taxifolin
Strongly bitter

Fig. 8.3 Stereochemistry influences the bitterness of molecules

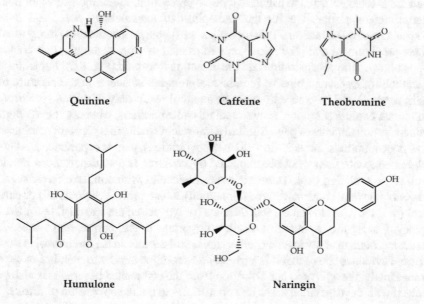

Quinine

Caffeine

Theobromine

Humulone

Naringin

Fig. 8.4 Structures of common bitter molecules

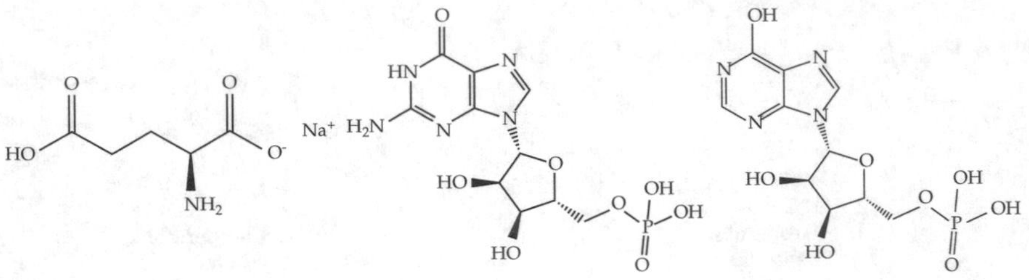

Fig. 8.5 Structures of umami-tasting molecules

taste" sensation and is the most frequently used acidifier in the food industry. Lactic and propionic acids are "cheesy", whereas acetic acid is typical of "vinegar". Malic acid is "green", whereas tartaric and phosphoric acids are described as "hard or intense".

Umami *taste* is the characteristic taste of salts of glutamic acid and is perceived through taste receptors that respond to the presence of glutamates. *Monosodium glutamate* (MSG), 5′-*ribonucleotides*, especially inosine 5′-monophosphate (IMP) and guanosine 5′-monophosphate (GMP) may all act synergistically, resulting in a unique umami sensation depending on the formulation (Fig. 8.5). MSG is naturally found in various foods (e.g., vegetables, meats, poultry, fish, shellfish, or aged cheese) and, together with ribonucleotides, are frequently used in snacks as flavour enhancers.

Other flavour-inducing molecules with a broad range of chemical structures exist and are associated with the trigeminal nerve perception. These molecules stimulate nerve signals responsible for the perception of *temperature*, *pain*, and *touch*. Capsaicin (chilli), zingerone (ginger), piperine (pepper) and menthol (mint) are the most predominant and all elicit signals translated as pain or temperature changes depending on the concentration and the detection threshold (Fig. 8.6). For example, menthol and capsaicin bind to protein receptors associated with temperature and pain perception. Low concentrations of menthol result only in odour. However, it induces a cooling sensation at intermediate concentrations, whereas at even higher concentrations, it causes pain. Similarly, capsaicin results in a pleasant spiciness at low concentrations, but at intermediate concentrations, pain is experienced. At even higher concentrations, heat accompanied by sweating is also experienced in addition to excruciating pain. There are many molecules with similar effects, such as eugenol (cloves), cinnamaldehyde (cinnamon), eucalyptol (eucalyptus oil) or carvacrol (oregano), just to name a few. Tannins (polyphenols) are responsible for astringency, that is, the dry-mouth sensation important in tea, coffee, and wines. The exact mechanism of astringency perception is still debated in the literature. However, current evidence shows that it is a trigeminal sensation similar to menthol or capsaicin. Finally, the *effervescence* sensation from the carbonated beverages is also perceived with the trigeminal nerve and contributes to the flavour profile of drinks.

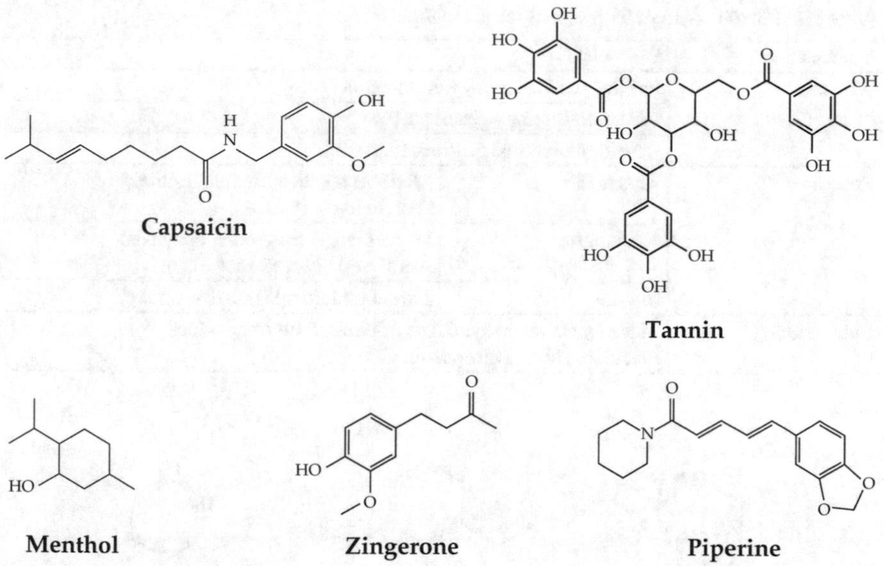

Fig. 8.6 Flavour-inducing molecules associated with the trigeminal nerve perception

8.3 Flavour Sources

Flavours may be generated from both natural and artificial sources (Table 8.1). *Plants* (e.g., elderberry flower) or *animals* (e.g., fish oil) can be used to *extract purified flavours* (see Sect. 8.1, example with vanilla) or make flavouring preparations. A *flavouring preparation* is a product of natural origin, which is not highly purified. For example, concentrated apple juice or yeast extracts can be defined as flavouring preparations. *Fermentation* can also be used to provide unique flavour characteristics to food that, in most cases, is impossible to replicate using another technology. The flavour profile of fermented foods is very complex depending on the starting raw material, cultures, and fermentation conditions. *Lactic acid* and *ethanol fermentations* produce a broad range of compounds that impart to the product unique flavour characteristics (e.g., diacetyl, acetaldehyde, or acetic acid). Fat and oil hydrolysis products also provide flavour in fermented foods (e.g., fruity esters or short and medium-chain fatty acids). *Process flavours* are generated mainly through Maillard reactions or caramelisation (Sect. 5.3). Finally, *smoke flavours* can be either obtained by collecting the smoke through distillation into a fluid (*liquid smoke*), which can then be applied in various foods (e.g., bacon) or used directly by introducing the smoke to the product from burning appropriate woods (e.g., hickory). The composition of smoke flavour depends on wood type and moisture content and most frequently consists of complex phenolic compounds, acids, carbonyl-containing compounds, and tars (Fig. 8.7). The smoke flavour is usually regulated

Table 8.1 Flavour sources with some typical examples

Sources	Examples	
Natural	Purified extracts (e.g., vanilla extract)	
	Concentrates (e.g., peach purée)	
	Raw ingredients (e.g., crushed garlic)	
Process	**Thermal**	Maillard reaction, caramelisation, Protein hydrolysates
	Enzymatic	Fermentation (microbial enzymes), Autolysed yeast extracts
	Smoke	Liquid or aerosol (i.e., real smoke)
Synthetic	Made by chemical synthesis of either food or non-food Precursors (e.g., petrochemicals).	

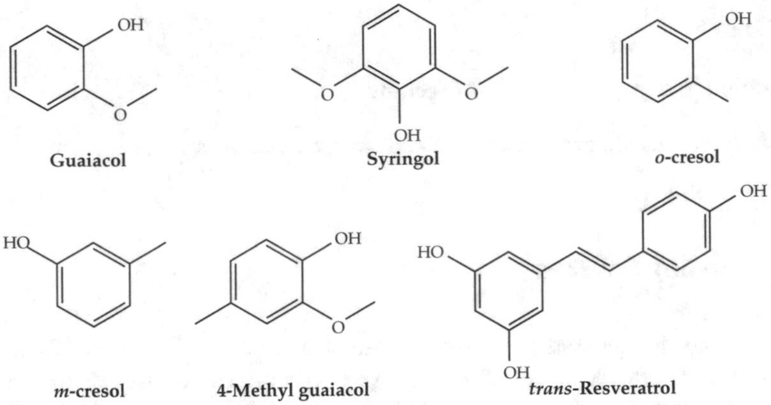

Fig. 8.7 Structures of molecules found in smoke flavour

separately from other flavours, as it contains unidentified substances with different safety requirements.

Plants belonging to the genus *Allium* (e.g., onion, garlic, shallot, leek, and chives) play a very important role in the generation of flavours of savoury food generating *organosulfur compounds* with strong flavours. These flavours play a central role in savoury-food manufacturing, and it is quite challenging to create such products without the inclusion of flavours from *Allium* species. These flavours may be either present in food formulations in the form of a raw ingredient (e.g., garlic paste) or very frequently as dehydrated powders (e.g., onion powder). Intact onions have very little odour, but cutting breaks cells and mixes *isoalliin* (substrate) with alliinase (enzyme), producing *sulphenic acids* and other products (Fig. 8.8). In addition, the enzyme *lachrymatory factor synthase* (LFS) converts sulphenic acids into a volatile molecule called *syn-propanethial S-oxide* responsible for tearing and the hot, pungent sensation of raw onion. Onion odours are derived from *spontaneous reactions*

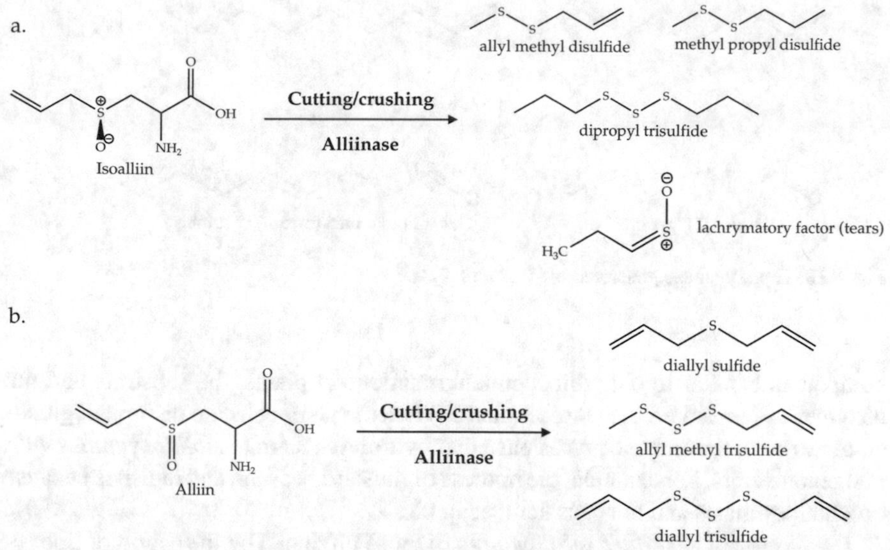

a.

allyl methyl disulfide   methyl propyl disulfide

Cutting/crushing

Alliinase

dipropyl trisulfide

Isoalliin

lachrymatory factor (tears)

b.

Cutting/crushing

Alliinase

Alliin

diallyl sulfide

allyl methyl trisulfide

diallyl trisulfide

Fig. 8.8 Flavour generation from plants of *Allium* species (**a**) onion, and (**b**) garlic

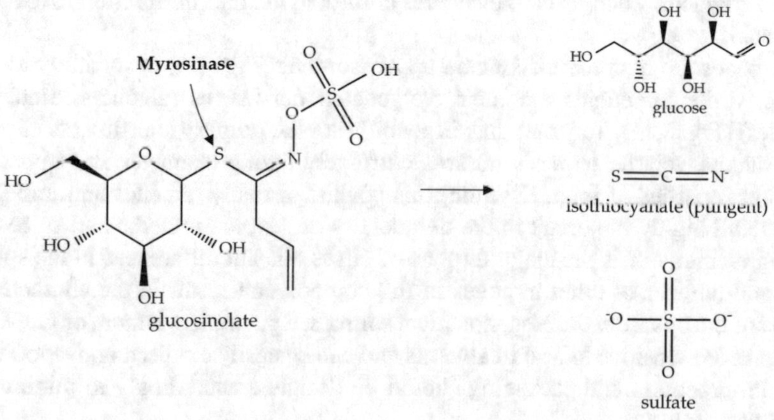

Myrosinase

glucose

isothiocyanate (pungent)

glucosinolate

sulfate

Fig. 8.9 Flavours derived from members of the Brassicaceae family

which convert sulphenic acids into thiosulphinate and other organosulfur compounds with flavour. Similar to onion, in garlic, the product of the reaction of alliin with alliinase yields allicin. Allicin is a very reactive molecule and is spontaneously broken down to other compounds such as *diallyl sulfide* and *diallyl disulfide*. Garlic flavours degrade rapidly with heat processing.

Glucosinolates are natural components of many plants belonging to the Brassicaceae family such as mustard, cabbage, or horseradish (Fig. 8.9). Glucosinolates are glycosides where the aglycone is a sulphur-containing

hexenal ("green")   nonadienal (melon/cucumber)

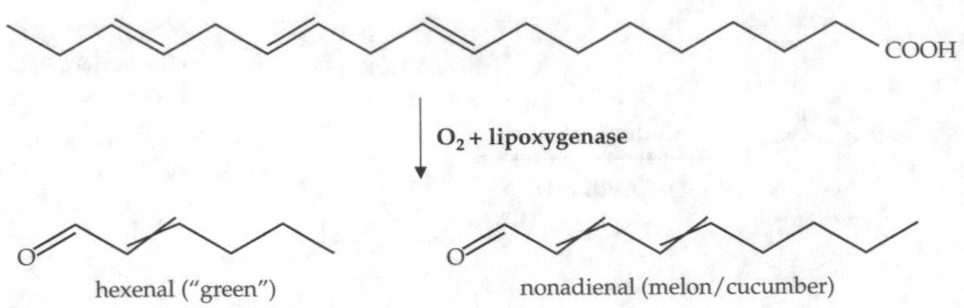

Fig. 8.10 Lipoxygenase generated off-flavours

compound. Similar to other biochemical reactions in plants, the substrate and the enzymes are separated in the intact plant. When the plant cells are disrupted, gluco-sinolates meet the enzyme myrosinase that hydrolyses them to *isothiocyanates* with pungent flavours. For example, the hotness of mustard, wasabi, and radish is because of isothiocyanates and not capsaicin, as in chilli.

Lipoxygenase generated off-flavours are those produced by the action of lipoxygenase on free polyunsaturated fatty acids (Fig. 8.10). Linoleic acid is the most common substrate in plant-based foods. Lipoxygenase has been associated with quality deterioration because of its involvement in off-flavour and odour production, loss of pigments such as carotenes and chlorophylls, and destruction of essential fatty acids.

Terpenes are hydrocarbons made up of isoprene (Fig. 7.6) occurring widely in plants, while terpenoids are their oxygenated derivatives (alcohols, aldehydes, ketones) (Fig. 8.11). Terpenes and terpenoids are the primary constituents of essential oils and volatile terpenes possess extremely strong character and impact the flavour properties of food. Essential oils consist of many terpenes and terpenoids that are all highly important in the uniqueness of flavours in herbs and spices. For example, limonene is the major terpene of citrus oils but differences in the specific type and number of other terpenes in the composition result in the characteristic aroma of the oils from different botanical sources, e.g., orange, lemon, or grapefruit. Terpenes are sensitive to heat treatments and can be easily oxidised or evaporate. As a result, processes and packaging should be designed accordingly to minimise or avoid flavour losses.

Protein hydrolysates are a complex mixture of peptides and amino acids produced by the partial hydrolysis of proteins. The name of a protein hydrolysate is specific to the ingredient. It includes the identity of the food source from which the protein was derived, for example, "hydrolysed wheat gluten", "hydrolysed soy protein", or "autolysed yeast extract". They are produced using heat and acid or enzymes and are most frequently used in snacks and savoury products to impart strong savoury sensations.

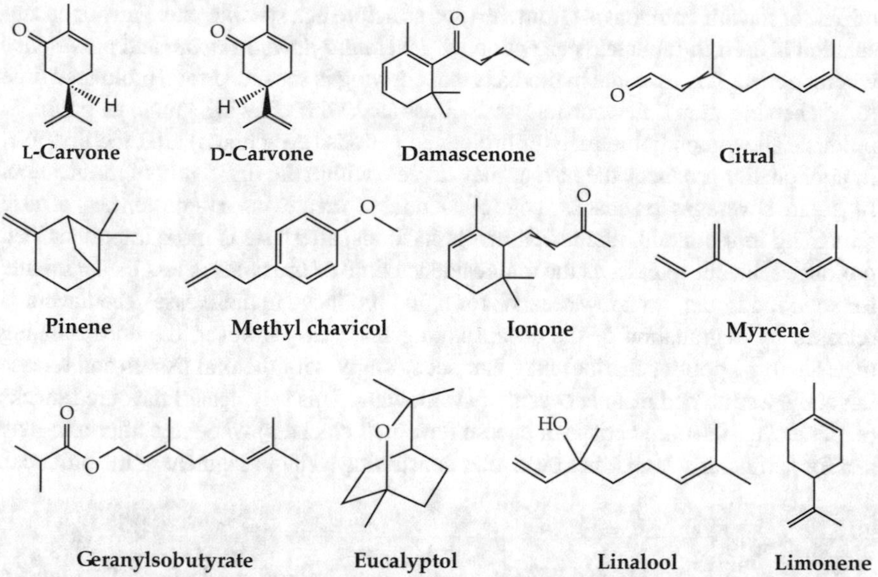

L-Carvone   D-Carvone   Damascenone   Citral

Pinene   Methyl chavicol   Ionone   Myrcene

Geranylsobutyrate   Eucalyptol   Linalool   Limonene

Fig. 8.11 Structures of common terpenes and terpenoids found in food

8.4 Flavour Delivery

Food composition and microstructure control flavour release and perception in foods, and both need to be taken into consideration when designing *delivery systems* for flavours. There are three ways to introduce a flavour into food. The flavouring can be added *as a solution* without the aid of any other process. For example, depending on its solubility, the flavour can be dissolved in vegetable oils or animal fats, water, ethanol, or glycerin and then introduced into the food. In beverages, flavour oils (e.g., orange oils) are *emulsified* most frequently with gum Arabic or modified starches. In this way, flavour oils that are not miscible with the aqueous phase of the formulation stay dispersed in the drink throughout the shelf life of the product. Flavours can also be *encapsulated* using various polysaccharides (e.g., maltodextrins or modified starches), proteins or lipids, and converted into a dry powder (e.g., spray drying). Encapsulation is quite similar in scope to emulsification, but it results in dry powder, whereas in emulsification, the protected flavour is in the liquid phase. Consequently, flavours may be dispersed uniformly in a dry form (e.g., powdered sauces). Encapsulation can protect flavours from other food ingredients and environmental conditions like heat, moisture, and acidity that cause degradation.

Flavours should also be released from the food in a controlled manner to avoid flavour saturation at the beginning of mastication and flavour depletion in the subsequent stages of oral processing. This process is termed *flavour release,* and it refers to the release of flavour components from the food structure at a specific rate. Flavour encapsulation is used to modulate perception by generating *flavour impact* and *persistence* where needed. It is essential in products that chewing is sustained for a prolonged time (e.g., chewing gum). Flavour release is illustrated with chewing gums, as premium products allow menthol release for prolonged periods (even hours) after the first bite. In poor quality products, the flavour may deplete within the first 5 min of mastication. In liquid beverages or snacks, controlled release creates an aftertaste, i.e., a taste remaining in the mouth after swallowing. In foods, aftertaste is more important than prolonged flavour release, as the oral residence time of the bolus is less than a minute for solid foods and just a few seconds for liquid products. In these cases, the flavour is released by degradation of the encapsulating material. However, the encapsulating material (e.g., polysaccharide) may interact strongly with the oral tissues and release flavour in a sustained manner ever after swallowing. This is typical of flavoured snacks (e.g., chicken-flavoured crisps or cheese flavoured crackers) where the aftertaste may last for particularly long times even after consuming a single segment of the product.

Encapsulation

Encapsulation is the process that entraps *active* components (e.g., flavours, vitamins, or drugs) within a carrier material known as the *wall*. The wall and the active form *supramolecular complexes* stabilised by non-covalent interactions (e.g., hydrogen, ionic, or hydrophobic interactions or a combination of them). The supramolecular complexes form particles that may be of micrometre (*microencapsulation*) or nanometre (*nanoencapsulation*) in size. Suitable wall materials for food applications are carbohydrates, proteins, or lipids. There are many encapsulation applications in food processing resulting in food with higher nutritional value, a lower dose of synthetic preservatives, and better sensory properties of the product. For instance, it can be used to enhance the stability of sensitive compounds during production, storage, ingestion (e.g., vitamins or drugs), decrease evaporation and degradation of volatile flavours, taste masking (e.g., astringency of polyphenols) or protect unsaturated fatty acids from oxygen, water, or light.

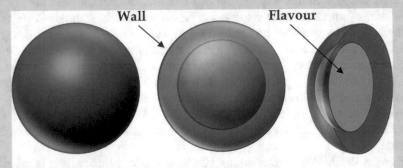

Intact, translucent and bisected views of an encapsulated flavour particle

8.5   Flavour Interactions and Stability

Flavour is influenced by food composition, as flavour molecules interact extensively with the rest of food components. Amorphous sugars (e.g., hard candy) provide a greater surface area for interaction than crystalline sugars (e.g., sugar crystals) while volatile flavours bind with varying degrees of strength to polysaccharides. For example, volatile flavouring substances can be entrapped in the helical structure of amylose, as the interior of the amylose helix is hydrophobic and can interact strongly with hydrophobic flavours. Starches with low amylose content (e.g., tapioca with ~17% amylose) and waxy starches consisting only of amylopectin have a weak binding capability. Those with a higher amylose content (like potato or maize) show greater interaction with flavours. Flavours also form strong bonds with native and denatured proteins. Proteins bind aldehydes and ketones through hydrophobic interactions, whereas alcohols are bound both hydrophobically and through hydrogen bonds. Most of the times, binding is reversible, so release during mastication is ensured. The binding capacity of fats is lower than that of oils. The quantity of bound flavour substance depends on the fatty acid chain-length and the presence of double bonds in the TAGs. TAGs with long-chain fatty acids bind less flavour than those with short-chain fatty acids, whereas unsaturated bind more flavour than saturated fatty acids.

Technical problems with flavours always need to be addressed to ensure consistent product quality and prolonged shelf life. Flavour preparations have stability problems, particularly in the presence of oxygen and light, as they may be easily oxidised and lose specific flavours. For example, certain plastic packaging materials have high permeability to O_2 or flavour molecules, may diffuse in or out of the product resulting in flavour changes during storage. Volatile components (e.g., aldehydes or ketones) may evaporate or flavour oils may physically separate (e.g., oil separation). Reactions with other food ingredients (e.g., Maillard reactions) may also occur, reducing the strength and may even alter the unique character of the flavour. For example, evaporation may occur during processing, such as in heat treatments or vacuum packaging. Ultimately, flavour choice depends on a combination of the factors mentioned above including cost and marketing considerations (e.g., natural flavour *vs* synthetic). Consequently, understanding the interactions between food components and flavour molecules is required to optimise formulation and extend the shelf life of products.

8.6   Learning Activities

8.6.1   Multiple-Choice Questions

1) A sweet-tasting molecule requires to satisfy the following conditions:
   a) To have a weak base, an electronegative group and a hydrophobic group
   b) To have a weak acid, an electronegative group and a hydrophobic group

c) To have a weak acid, a hydroxyl group and a hydrophobic group

d) To have a weak acid, a hydroxyl group and an amino group

2) Bitter tasting molecules have:

a) One polar and one non-polar group in their structure

b) One polar and two non-polar groups in their structure

c) Two polar and two non-polar groups in their structure

d) Two non-polar groups in their structure

3) Bitter tasting molecules are structurally diverse, mostly being:

a) Alkaloids, glycosides, and peptides

b) Glycosides, ketones, and acids

c) Alkaloids, amino acids, and aldehydes

d) Peptides, acids, and xanthines

4) Alliinase in onions catalyses the conversion of:

a) Isoalliin to thiosulfinate

b) Alliin to thiosulfinate

c) Isoallicin to thiosulfinate

d) Alicin to thiosulfonate

5) Myrosinase is the enzyme responsible for flavour generation in:

a) Allium family

b) Brassicaceae family

c) In both Allium and Brassicaceae families

d) Nutmeg to convert myristic acid to aldehydes with odour

6) The stereochemistry (i.e., D-, L- configurations) may change how a molecule smells. This statement is:

a) True

b) False

7) Most essential oils consist of limonene, pinene and myrcene in different proportions resulting in different smells. This statement is:

a) True

b) False

8) Smoke flavours are chemically

a) Phenolic compounds

b) Aldehydes

c) Terpenes

d) Pinenes

9) Quinine tastes

   a) Sweet
   b) Salty
   c) Bitter
   d) Umami

10) Guanosine 5′-monophosphate tastes

   a) Bitter
   b) Salty
   c) Sweet
   d) Umami

8.6.2 Short Answer Questions – Further Reading

1) **Online activity** Search online the sweeteners thaumatin, steviol glycoside, sucralose and neotame and compare their structures. Can you identify possible AH^+, B^-, and γ regions in the structures?

2) **Online activity** Search online for the principles of flavour encapsulation technology and its uses in the food industry.

3) Discuss the molecular interactions of carbohydrates, proteins, and lipids with flavour compounds.

4) Discuss technical problems of flavours that relate to their chemical and physical stability.

5) A food product contains "yeast extract" and "malt extract". What is the role of these ingredients in the formulation? How are they produced, and what do they most likely taste?

6) Draw the structure of limonene, α-pinene, citral, geraniol, and linalool. What is the difference between terpenes and terpenoids? Which are terpenes or terpenoids from the structures you have just drawn? Identify the isoprene chain in citral, geraniol, and linalool. How many isoprene units can you count?

7) **Further reading** Find the article "Ley, J. P. (2008). Masking Bitter Taste by Molecules. *Chemosensory Perception*, 1(1), 58–77."

   • Bitter masking can be achieved by "congruent flavours", "complexing agents", and "low molecular weight substances". Name compounds that can be used in each technology.

8) **Further reading** Find the article "Gharsallaoui, A., Roudaut, G., Chambin, O., Voilley, A., & Saurel, R. (2007). Applications of spray-drying in microencapsulation of food ingredients: An overview. *Food Research International*, 40(9), 1107–1121."

- Describe the three basic steps taken in encapsulation of food ingredients?
- What are the most used wall materials in the encapsulation of food compounds? Name two from each class.
- Give five examples of ingredients encapsulated with spray drying

8.6.3   Fill the Gaps

1) A sweet molecule must have the correct molecular _____ and the proper _____ distribution.

2) Quinine is a _____ alkaloid whereas caffeine is a _____ alkaloid.

3) Monosodium glutamate, inosine 5′-monophosphate and guanosine 5′-monophosphate all taste _____.

4) Spicy, cool, or astringent are sensations associated with the _____ nerve perception.

5) Liquid smoke is obtained through _____.

6) It is difficult to create savoury products without inclusion of flavours from _____ species.

7) _____ are natural components of many plants belonging to the Brassicaceae family such as

8) _____ may generate off flavours with its action on free polyunsaturated fatty acids

9) _____ and _____ are the primary constituents of essential oils.

10) Terpenes are sensitive to heat treatments and can be easily _____ or _____.

Appendices

Chapter 1 – Water

Multiple choice questions: 1d, 2c, 3d, 4b, 5a, 6b, 7c, 8a, 9c,10a
Fill the gaps: 1) polarity, polar, 2) hydrogen bonding, 3) hydrophobic, 4) osmosis, 5) intensity/strength, 6) 0.5, 0.8, 7) molecular mobility, highest, 8) increases, 9) glass transition, 10) plasticisation, plasticiser.

Chapter 2 – Carbohydrates

Multiple choice questions – Monosaccharides: 1d, 2a, 3a, 4a, 5b, 6c, 7b, 8a, 9a, 10b
Multiple choice questions – Polysaccharides: 1a, 2c, 3a, 4d, 5a, 6a, 7b, 8c, 9c, 10c
Fill the gaps: 1) four, C*, 2) enantiomers, mirror images, 3) hemiacetals, 4) below, above, 5) chair, boat, 6) uronic acids, 7) reducing sugars, 8) linear, branched, 9) formation, stabilisation, 10) A-type, B-type, 11) chemical, physical, functional, 12) recrystallisation, amylose, amylopectin, 13) 100, 0, 14) Carrageenan, 15) homogalacturonan, rhamnogalacturonan-I, 16) high methoxy, low methoxy, 17) flavour, 18) low, 19) chitin, 20) hemicellulose

Chapter 3 – Proteins

Multiple choice questions – Proteins: 1a, 2b, 3b, 4b, 5c, 6a, 7a, 8c, 9b, 10a
Multiple choice questions – Enzymes: 1a, 2b, 3c, 4d, 5a, 6b, 7c, 8d, 9a, 10a

© The Author(s), under exclusive license to Springer Nature Switzerland AG 2021
V. Kontogiorgos, *Introduction to Food Chemistry*,
https://doi.org/10.1007/978-3-030-85642-7

Fill the gaps: 1) Essential, 2) pH, zero, 3) cysteines, cystine, 4) disulfide bond, 5) carbon–nitrogen, 6) ninhydrin, 7) Legumin, vicilin, 8) folding, 3D structure, 9) fibrous, globular, 10) physical, chemical, 11) hydrophobic interactions, 12) lyotropic, 13) film, oil, 14) activation energy, 15) specificity, 16) isoenzymes, 17) metalloenzymes, 18) low, 19) main class, 20) Endoacting, 21) aminopeptidases, carboxypeptidases, 22) phenolic, 23) transglutaminases, 24) debittering

Chapter 4 – Lipids

Multiple choice questions: 1b, 2b, 3d, 4a, 5d, 6c, 7a, 8a, 9a, 10a
Fill the gaps: 1) length, saturation, 2) double, 3) cis-, 4) single, 5) position, 6) phytosterols, 7) auto-oxidation, photo-oxidation, enzymatic oxidation, 8) hydroperoxides, 9) a<β′<β, 10) polymorphic, transition, 11) tempered, 12) solid fat content, 13) catalysts, enzymes

Chapter 5 – Browning

Multiple choice questions: 1a, 2b, 3c, 4b, 5d, 6a, 7a, 8c, 9a, 10b
Fill the gaps: 1) *p*-diphenols, 2) carapace, 3) ascorbic, sulphites, 4) thermal degradation, 5) furfural, maltol, 6) ammonium, sulfite, 7) ketosamines, 8) amino, Strecker, 9) melanoidins, melanins, 10) asparagine

Chapter 6 – Vitamins-Minerals

Multiple choice questions: 1a, 2a, 3a, 4b, 5d, 6c, 7a, 8a, 9d, 10a
Fill the gaps: 1) dehydrated, 2) Vitamin D, 3) thiamine, 4) heat, acidic, 5) Enrichment, 6) solubility, 7) chelates, 8) Phytic, 9) polyphenols, tannins, 10) bioavailability

Chapter 7 – Colour chemistry

Multiple choice questions: 1a, 2a, 3a, 4b, 5b, 6a, 7b, 8a, 9b, 10a
Fill the gaps: 1) quinine, 2) cutin, 3) oxidation, heme, 4) brown, 5) pink, 6) magnesium, pheophytin, pyropheophytin, 7) isoprene, conjugated, 8) trans, cis, 9) anthocyanidins, 10) blue/violet, 11) Betacyanins, 12) azo

Chapter 8 – Flavour chemistry

Multiple choice questions: 1b, 2a, 3a, 4a, 5b, 6a, 7b, 8a, 9c, 10d

Fill the gaps: 1) shape, electronic, 2) quinidine, xanthine, 3) umami, 4) trigeminal, 5) distillation, 6) Allium, 7) Glucosinolates, 8) Lipoxygenase, 9) terpenes, terpenoids,10) oxidise, evaporate

Bibliography

Books

Ahmed J (2017) Glass Transition and Phase Transitions in Food and Biological Materials. John Wiley & Sons, Chichester, UK

Barbosa-Cánovas GV, Fontana AJ, Schmidt SJ, Labuza TP (2020) Water Activity in Foods: Fundamentals and Applications. Wiley,

Belitz HD, Grosch W, Schieberle P (2009) Food chemistry. 4th edn. Springer, Berlin

BeMiller JN, Whistler RL (2009) Starch: Chemistry and Technology. ISSN. Elsevier Science,

Brady WJ (2013) Introductory food chemistry. Cornell University Press, New York

Casimir CA (2017) Food Lipids: Chemistry, Nutrition, and Biotechnology. 4th edn. CRC Press, Boca Raton

Coultate TP (2016) Food: the chemistry of its components. Food, the chemistry of its components, 6th edition. edn. Royal Society of Chemistry, Cambridge, UK

Coupland J (2014) An Introduction to the Physical Chemistry of Food. Food Science Text Series. Springer-Verlag, New York.

Cui SW (2005) Food Carbohydrates: Chemistry, Physical Properties, and Applications. CRC Press,

Damodaran S, Parkin KL (2017) Fennema's Food Chemistry 5th edn. CRC Press.

DeMan JM, Finley JW, Hurst WJ, Lee CY (2018) Principles of food chemistry. 4th edn. Springer, Switzerland

Emerton V (2008) Food colours. Leatherhead Pub. Blackwell Pub., Surrey, UK : Oxford, UK

Guichard E, Salles C, Morzel M, Le Bon A-M (2017) Flavour: From Food to Perception. John Wiley & Sons, Chichester, UK.

Melton LD, Shahidi F, Varelis P (2019) Encyclopedia of food chemistry. Elsevier, Amsterdam, Netherlands

Parisi S, Ameen SM, Montalto S, Santangelo A (2019) Maillard reaction in foods : mitigation strategies and positive properties. Springer, Cham

Phillips OG, Williams AP (2011) Handbook of Food Proteins. Woodhead Publishing, Oxford, UK

Phillips OG, Williams AP (eds) (2020) Handbook of Hydrocolloids. 3rd edn. Elsevier.

Socaciu C (2008) Food colorants : chemical and functional properties. CRC Press, Boca Raton

Tuvikene R (2021) Chapter 25 – Carrageenans. In: Phillips GO, Williams PA (eds) Handbook of Hydrocolloids (Third Edition). Woodhead Publishing, pp 767–804.

Velíšek J, Koplik R, Cejpek K (2020) The chemistry of food. 2nd edn. Wiley Blackwell, Chichester, UK

© The Author(s), under exclusive license to Springer Nature Switzerland AG 2021   195
V. Kontogiorgos, *Introduction to Food Chemistry*,
https://doi.org/10.1007/978-3-030-85642-7

Voilley Ae, Etivéant P (2006) Flavour in food. CRC Press Woodhead Publishing, Boca Raton, FL : Cambridge, England

Wang D (2012) Food chemistry. Nova Science Publishers, Hauppauge, N.Y.

Whitaker JR, Voragen AGJ, Wong DWS (2003) Handbook of food enzymology. Marcel Dekker, New York

Wong DWS (2018) Mechanism and theory in food chemistry. 2nd edn. Springer, Cham, Switzerland

Zeece M (2020) Introduction to the chemistry of food. Academic Press, London

Review articles

Azeredo HMC (2009) Betalains: properties, sources, applications, and stability – a review. International Journal of Food Science & Technology 44 (12):2365–2376. doi:https://doi.org/10.1111/j.1365-2621.2007.01668.x

Barden L, Decker EA (2016) Lipid Oxidation in Low-moisture Food: A Review. Critical Reviews in Food Science and Nutrition 56 (15):2467–2482. doi:https://doi.org/10.1080/10408398.2013.848833

Barriuso B, Astiasarán I, Ansorena D (2013) A review of analytical methods measuring lipid oxidation status in foods: a challenging task. European Food Research and Technology 236 (1):1–15. doi:https://doi.org/10.1007/s00217-012-1866-9

Boye J, Zare F, Pletch A (2010) Pulse proteins: Processing, characterization, functional properties and applications in food and feed. Food Research International 43 (2):414–431. doi:https://doi.org/10.1016/j.foodres.2009.09.003

Cao L, Lu W, Mata A, Nishinari K, Fang Y (2020) Egg-box model-based gelation of alginate and pectin: A review. Carbohydrate Polymers 242:116389. doi:https://doi.org/10.1016/j.carbpol.2020.116389

Capuano E, Fogliano V (2011) Acrylamide and 5-hydroxymethylfurfural (HMF): A review on metabolism, toxicity, occurrence in food and mitigation strategies. LWT – Food Science and Technology 44 (4):793–810. doi:https://doi.org/10.1016/j.lwt.2010.11.002

Carocho M, Barreiro MF, Morales P, Ferreira ICFR (2014) Adding Molecules to Food, Pros and Cons: A Review on Synthetic and Natural Food Additives. Comprehensive Reviews in Food Science and Food Safety 13 (4):377–399. doi:https://doi.org/10.1111/1541-4337.12065

Carocho M, Morales P, Ferreira ICFR (2017) Sweeteners as food additives in the XXI century: A review of what is known, and what is to come. Food and Chemical Toxicology 107:302–317. doi:https://doi.org/10.1016/j.fct.2017.06.046

Cavalcanti RN, Santos DT, Meireles MAA (2011) Non-thermal stabilization mechanisms of anthocyanins in model and food systems—An overview. Food Research International 44 (2):499–509. doi:https://doi.org/10.1016/j.foodres.2010.12.007

Chang C, Wu G, Zhang H, Jin Q, Wang X (2020) Deep-fried flavor: characteristics, formation mechanisms, and influencing factors. Critical Reviews in Food Science and Nutrition 60 (9):1496–1514. doi:https://doi.org/10.1080/10408398.2019.1575792

Chen B, McClements DJ, Decker EA (2011) Minor Components in Food Oils: A Critical Review of their Roles on Lipid Oxidation Chemistry in Bulk Oils and Emulsions. Critical Reviews in Food Science and Nutrition 51 (10):901–916. doi:https://doi.org/10.1080/10408398.2011.606379

Choe E, Min DB (2006) Mechanisms and Factors for Edible Oil Oxidation. Comprehensive Reviews in Food Science and Food Safety 5 (4):169-186. doi:https://doi.org/10.1111/j.1541-4337.2006.00009.x

Day L (2013) Proteins from land plants – Potential resources for human nutrition and food security. Trends in Food Science & Technology 32 (1):25–42. doi:https://doi.org/10.1016/j.tifs.2013.05.005

de Oliveira FC, Coimbra JSdR, de Oliveira EB, Zuñiga ADG, Rojas EEG (2016) Food Protein-polysaccharide Conjugates Obtained via the Maillard Reaction: A Review. Critical Reviews in Food Science and Nutrition 56 (7):1108–1125. doi:https://doi.org/10.1080/1040839 8.2012.755669

Dickinson E (2003) Hydrocolloids at interfaces and the influence on the properties of dispersed systems. Food Hydrocolloids 17 (1):25–39. doi:http://dx.doi.org/10.1016/S0268-005X(01)00120-5

Dickinson E (2018) Hydrocolloids acting as emulsifying agents – How do they do it? Food Hydrocolloids 78:2–14. doi:j.foodhyd.2017.01.025

Elleuch M, Bedigian D, Roiseux O, Besbes S, Blecker C, Attia H (2011) Dietary fibre and fibre-rich by-products of food processing: Characterisation, technological functionality and commercial applications: A review. Food Chemistry 124 (2):411–421. doi:https://doi.org/10.1016/j.foodchem.2010.06.077

Fang Z, Bhandari B (2010) Encapsulation of polyphenols – a review. Trends in Food Science & Technology 21 (10):510–523. doi:https://doi.org/10.1016/j.tifs.2010.08.003

Fathi M, Martín Á, McClements DJ (2014) Nanoencapsulation of food ingredients using carbohydrate based delivery systems. Trends in Food Science & Technology 39 (1):18–39. doi:https://doi.org/10.1016/j.tifs.2014.06.007

Finley JW, Kong A-N, Hintze KJ, Jeffery EH, Ji LL, Lei XG (2011) Antioxidants in Foods: State of the Science Important to the Food Industry. Journal of Agricultural and Food Chemistry 59 (13):6837–6846. doi:https://doi.org/10.1021/jf2013875

Foegeding EA (2006) Food Biophysics of Protein Gels: A Challenge of Nano and Macroscopic Proportions. Food Biophysics 1 (1):41–50. doi:https://doi.org/10.1007/s11483-005-9003-y

Foegeding EA, Davis JP (2011) Food protein functionality: A comprehensive approach. Food Hydrocolloids 25 (8):1853–1864. doi:https://doi.org/10.1016/j.foodhyd.2011.05.008

Fuentes-Zaragoza E, Riquelme-Navarrete MJ, Sánchez-Zapata E, Pérez-Álvarez JA (2010) Resistant starch as functional ingredient: A review. Food Research International 43 (4):931–942. doi:https://doi.org/10.1016/j.foodres.2010.02.004

Gharsallaoui A, Roudaut G, Chambin O, Voilley A, Saurel R (2007) Applications of spray-drying in microencapsulation of food ingredients: An overview. Food Research International 40 (9):1107–1121. doi:https://doi.org/10.1016/j.foodres.2007.07.004

Gómez-Guillén MC, Giménez B, López-Caballero ME, Montero MP (2011) Functional and bioactive properties of collagen and gelatin from alternative sources: A review. Food Hydrocolloids 25 (8):1813–1827. doi:https://doi.org/10.1016/j.foodhyd.2011.02.007

Gouin S (2004) Microencapsulation: industrial appraisal of existing technologies and trends. Trends in Food Science & Technology 15 (7):330–347. doi:https://doi.org/10.1016/j.tifs.2003.10.005

Huff Lonergan E, Zhang W, Lonergan SM (2010) Biochemistry of postmortem muscle – Lessons on mechanisms of meat tenderization. Meat Science 86 (1):184-195. doi:https://doi.org/10.1016/j.meatsci.2010.05.004

Hunt JR (2003) Bioavailability of iron, zinc, and other trace minerals from vegetarian diets. The American Journal of Clinical Nutrition 78 (3):633S-639S. doi:https://doi.org/10.1093/ajcn/78.3.633S

Khan MI (2016) Stabilization of betalains: A review. Food Chemistry 197:1280-1285. doi:https://doi.org/10.1016/j.foodchem.2015.11.043

Kiani H, Sun D-W (2011) Water crystallization and its importance to freezing of foods: A review. Trends in Food Science & Technology 22 (8):407-426. doi:https://doi.org/10.1016/j.tifs.2011.04.011

Kumar V, Sinha AK, Makkar HPS, Becker K (2010) Dietary roles of phytate and phytase in human nutrition: A review. Food Chemistry 120 (4):945–959. doi:https://doi.org/10.1016/j.foodchem.2009.11.052

Lam RSH, Nickerson MT (2013) Food proteins: A review on their emulsifying properties using a structure–function approach. Food Chemistry 141 (2):975–984. doi:https://doi.org/10.1016/j.foodchem.2013.04.038

Leopoldini M, Russo N, Toscano M (2011) The molecular basis of working mechanism of natural polyphenolic antioxidants. Food Chemistry 125 (2):288–306. doi:https://doi.org/10.1016/j.foodchem.2010.08.012

Ley JP (2008) Masking Bitter Taste by Molecules. Chemosensory Perception 1 (1):58–77. doi:https://doi.org/10.1007/s12078-008-9008-2

Lineback DR, Coughlin JR, Stadler RH (2012) Acrylamide in Foods: A Review of the Science and Future Considerations. Annual Review of Food Science and Technology 3 (1):15–35. doi:https://doi.org/10.1146/annurev-food-022811-101114

Lucey JA (2002) Formation and Physical Properties of Milk Protein Gels. Journal of Dairy Science 85 (2):281–294. doi:https://doi.org/10.3168/jds.S0022-0302(02)74078-2

Lund MN, Ray CA (2017) Control of Maillard Reactions in Foods: Strategies and Chemical Mechanisms. Journal of Agricultural and Food Chemistry 65 (23):4537–4552. doi:https://doi.org/10.1021/acs.jafc.7b00882

Marangoni AG, Acevedo N, Maleky F, Co E, Peyronel F, Mazzanti G, Quinn B, Pink D (2012) Structure and functionality of edible fats. Soft Matter 8 (5):1275–1300. doi:https://doi.org/10.1039/C1SM06234D

Marangoni AG, van Duynhoven JPM, Acevedo NC, Nicholson RA, Patel AR (2020) Advances in our understanding of the structure and functionality of edible fats and fat mimetics. Soft Matter 16 (2):289–306. doi:https://doi.org/10.1039/C9SM01704F

Marcus Y (2009) Effect of Ions on the Structure of Water: Structure Making and Breaking. Chemical Reviews 109 (3):1346–1370. doi:https://doi.org/10.1021/cr8003828

McClements DJ (2004) Protein-stabilized emulsions. Current Opinion in Colloid & Interface Science 9 (5):305–313. doi:https://doi.org/10.1016/j.cocis.2004.09.003

McClements DJ, Rao J (2011) Food-Grade Nanoemulsions: Formulation, Fabrication, Properties, Performance, Biological Fate, and Potential Toxicity. Critical Reviews in Food Science and Nutrition 51 (4):285–330. doi:https://doi.org/10.1080/10408398.2011.559558

Ngamwonglumlert L, Devahastin S, Chiewchan N (2017) Natural colorants: Pigment stability and extraction yield enhancement via utilization of appropriate pretreatment and extraction methods. Critical Reviews in Food Science and Nutrition 57 (15):3243–3259. doi:https://doi.org/10.1080/10408398.2015.1109498

Nishinari K, Fang Y, Guo S, Phillips GO (2014) Soy proteins: A review on composition, aggregation and emulsification. Food Hydrocolloids 39:301–318. doi:https://doi.org/10.1016/j.foodhyd.2014.01.013

Pathare PB, Opara UL, Al-Said FA-J (2013) Colour Measurement and Analysis in Fresh and Processed Foods: A Review. Food and Bioprocess Technology 6 (1):36–60. doi:https://doi.org/10.1007/s11947-012-0867-9

Patras A, Brunton NP, O'Donnell C, Tiwari BK (2010) Effect of thermal processing on anthocyanin stability in foods; mechanisms and kinetics of degradation. Trends in Food Science & Technology 21 (1):3–11. doi:https://doi.org/10.1016/j.tifs.2009.07.004

Pérez S, Bertoft E (2010) The molecular structures of starch components and their contribution to the architecture of starch granules: A comprehensive review. Starch - Stärke 62 (8):389–420. doi:https://doi.org/10.1002/star.201000013

Post MJ (2012) Cultured meat from stem cells: Challenges and prospects. Meat Science 92 (3):297–301. doi:https://doi.org/10.1016/j.meatsci.2012.04.008

Purlis E (2010) Browning development in bakery products – A review. Journal of Food Engineering 99 (3):239–249. doi:https://doi.org/10.1016/j.jfoodeng.2010.03.008

Roos YH (2010) Glass Transition Temperature and Its Relevance in Food Processing. Annual Review of Food Science and Technology 1 (1):469–496. doi:https://doi.org/10.1146/annurev.food.102308.124139

Saha D, Bhattacharya S (2010) Hydrocolloids as thickening and gelling agents in food: a critical review. J Food Sci Technol 47 (6):587–597. doi:https://doi.org/10.1007/s13197-010-0162-6

Saini RK, Nile SH, Park SW (2015) Carotenoids from fruits and vegetables: Chemistry, analysis, occurrence, bioavailability and biological activities. Food Research International 76:735–750. doi:https://doi.org/10.1016/j.foodres.2015.07.047

Schmid A, Walther B (2013) Natural Vitamin D Content in Animal Products. Advances in Nutrition 4 (4):453–462. doi:https://doi.org/10.3945/an.113.003780

Singh A, Auzanneau FI, Rogers MA (2017) Advances in edible oleogel technologies – A decade in review. Food Research International 97:307–317. doi:https://doi.org/10.1016/j.foodres.2017.04.022

Stintzing FC, Carle R (2004) Functional properties of anthocyanins and betalains in plants, food, and in human nutrition. Trends in Food Science & Technology 15 (1):19–38. doi:https://doi.org/10.1016/j.tifs.2003.07.004

Totosaus A, Montejano JG, Salazar JA, Guerrero I (2002) A review of physical and chemical protein-gel induction. International Journal of Food Science & Technology 37 (6):589–601. doi:https://doi.org/10.1046/j.1365-2621.2002.00623.x

Turek C, Stintzing FC (2013) Stability of Essential Oils: A Review. Comprehensive Reviews in Food Science and Food Safety 12 (1):40–53. doi:https://doi.org/10.1111/1541-4337.12006

van Boekel M, Fogliano V, Pellegrini N, Stanton C, Scholz G, Lalljie S, Somoza V, Knorr D, Jasti PR, Eisenbrand G (2010) A review on the beneficial aspects of food processing. Molecular Nutrition & Food Research 54 (9):1215–1247. doi:https://doi.org/10.1002/mnfr.200900608

Varela P, Fiszman SM (2011) Hydrocolloids in fried foods. A review. Food Hydrocolloids 25 (8):1801–1812. doi:https://doi.org/10.1016/j.foodhyd.2011.01.016

Ventura Alison K, Worobey J (2013) Early Influences on the Development of Food Preferences. Current Biology 23 (9):R401–R408. doi:https://doi.org/10.1016/j.cub.2013.02.037

Wang FC, Gravelle AJ, Blake AI, Marangoni AG (2016) Novel trans fat replacement strategies. Current Opinion in Food Science 7:27–34. doi:https://doi.org/10.1016/j.cofs.2015.08.006

Wang H-Y, Qian H, Yao W-R (2011) Melanoidins produced by the Maillard reaction: Structure and biological activity. Food Chemistry 128 (3):573–584. doi:https://doi.org/10.1016/j.foodchem.2011.03.075

Waraho T, McClements DJ, Decker EA (2011) Mechanisms of lipid oxidation in food dispersions. Trends in Food Science & Technology 22 (1):3–13. doi:https://doi.org/10.1016/j.tifs.2010.11.003

Yoruk R, Marshall MR (2003) Physicochemical properties and function of plant polyphenol oxidase: a review. Journal of Food Biochemistry 27 (5):361–422. doi:https://doi.org/10.1111/j.1745-4514.2003.tb00289.x

Index

A

Acacia gum, 44
Acetal, 22
Acrylamide, 132
Activation energy, 74
Active site, 75
Aftertaste, 186
Aglycone, 28
Albumins, 62
Aldaric, 27
Alditol, 28
Aldonic, 27
Aldoses, 20
Alginate, 40
Alkaloids, 178
Alkyl radical, 105
Allicin, 183
Alliinase, 182
Allosteric inhibition, 81
Alternative protein, 56
Amadori compounds, 132
Amadori rearrangement, 132
Amino acid, 56
Aminopeptidases, 86
Amorphous, 110
Amphiphilic, 102
Amphoteric, 58
Amylases, 83
Amyloglucosidase, 84
Amylopectin, 34
Amylose, 34
Anomeric carbon, 23
Anthocyanidins, 165
Anthocyanins, 165
Antioxidants, 108

Apo-enzyme, 76
Artificial flavours, 175
Ascorbic acid, 144
Astringency, 180
Auto-oxidation, 105
Axial, 24
Azo-dyes, 169

B

Bacteriocins, 61
Betacyanins, 167
Betalains, 167
Betalamic acid, 167
Betaxanthins, 167
Bioactive peptides, 60
Biocytin, 148
Biotin, 148
Bitter, 177
Bitter peptides, 60
Blooming, 115
Boat, 24

C

Caramel colours, 128
Caramel flavours, 128
Caramelisation, 128
Carboxypeptidases, 86
Carotenes, 163
Carotenoids, 163
Carrageenan, 39
Catalyst, 74
Cellulases, 84
Cellulose, 43

© The Author(s), under exclusive license to Springer Nature Switzerland AG 2021
V. Kontogiorgos, *Introduction to Food Chemistry*,
https://doi.org/10.1007/978-3-030-85642-7

Cellulose derivatives, 43
Ceramides, 103
Cerebroside, 103
Chair, 24
Chelators, 108
Chiral carbon, 20
Chitin, 46
Chitosan, 46
Chlorophyll, 162
Chlorophyllides, 163
Cholecalciferol, 143
Cholesterol, 104
Chymosin, 86
cis, 99
Coagulation, 67
Cobalamin, 147
Co-enzyme, 76
Co-factor, 76
Collagen, 62
Colligative properties, 6
Colourant, 160
Competitive inhibition, 81
Conformation, 24
Conformers, 24
Conjugated double bond, 100
Conjugated linoleic acid, 100
Crystal, 110
Crystallisation, 111
Cystine, 61

D
Degree of esterification (DE), 42
Denaturation, 67, 69
Dextrins, 83
Dextrose equivalents (DE), 39
Diacylglycerol, 101
Diastereoisomerism, 20
Dietary fibre (DF), 43, 46
Dipolar ion, 59
Disulfide bond, 60, 61
Disulfide bridge, 61
Dye, 160

E
EC number, 82
Egg-box model, 40
Emulsion, 73
Enantiomer, 20
Enantiomerism, 20
Enantiomers, 21
Endoacting enzymes, 82
Enol, 26

Enzyme affinity, 78
Enzyme inhibitors, 80
Enzyme kinetics, 77
Epimers, 20, 24
Equatorial, 24
Ergocalciferol, 143
Esterification, 26, 101
Exoacting enzymes, 82

F
Fat, 97
Fatty acids, 98
Fischer projection, 22
Flavonoids, 167
Flavour, 175
Flavylium cation, 165
Foam, 73
Folding, 63
Folic acid, 147
Fructans, 32
Furanose, 23
Furfural, 128

G
Galactomannan, 43
Ganglioside, 103
Gel, 38, 67, 72
Gelatin, 62
Gelatinisation, 36, 37
Gelation, 37, 67
Gliadin, 62
Globulins, 62
Glucoamylase, 84
Glucono delta-lactone (GdL), 27
Glucose isomerase, 84
Glucose syrup, 39, 84
Glucosinolates, 183
Glutelins, 62
Gluten, 38, 61, 62
Gluten-free, 38
Glutenin, 62
Glycosaminoglycans, 46
Glycoside, 28
Glycosidic bond, 28
Growth rings, 36
Gum Arabic, 44

H
Haem, 161
Hairy region, 42
Haworth projection, 23

Helix, 64
Hemiacetal, 22
Hemi-celluloses, 46
Hexagonal ice, 2
High amylose, 36
High fructose corn syrup (HFCS), 84
High methoxy pectins (HM-pectin), 42
Hilum, 36, 37
Hofmeister series, 70
Holo-enzyme, 76
Homogalacturonan (HG), 42
Hydrogenation, 116
Hydrogen bond, 3
Hydrolytic rancidity, 87, 109
Hydroperoxides (ROOH), 105
Hydrophobic interactions, 5
Hydroxymethyl furfural (HMF), 128
Hypertonic, 6
Hypotonic, 6

I
Imine, 61
Initial velocity (V_o), 77
Interesterification, 118
Inulin, 32
Invertase, 86
Invert sugar, 31, 86
Ionic interactions, 4
Iso-amylases, 84
Isoelectric point (pI), 59
Isoenzyme, 74
Isoprene rule, 163

K
Ketosamines, 132
Ketoses, 20

L
Laccase, 125
Lactase, 31, 86
Lactones, 27
Lactose, 31
Lactose free, 86
Lake, 160
Lecithin, 102
Ligands, 150
Lignin, 46
Lipases, 87
Lipid oxidation, 105
Lipids, 97

Lipolysis, 109
Lipoxygenases, 88, 184
Loop, 65
Low methoxy pectins (LM-pectin), 42
Lyotropic series, 70

M
Maillard reaction, 61, 130
Maltodextrin, 39
Maltol, 128
Maltose, 31
Maltose syrup, 84
Margarine, 104, 111
Maximum velocity (V_{max}), 77
Melanoidins, 134
Melting point, 115
Metal-activated enzymes, 76
Metalloenzyme, 76
Metmyoglobin, 161
Michaelis-Menten constant, 78
Modified starch, 27, 34, 36, 38
Moisture sorption isotherm, 9
Molecular mobility, 9
Monoacylglycerol, 101
Monomolecular layer, 9
Monosaccharide, 20
Monounsaturated, 98
Motif, 65
Mutarotation, 25
Myoglobin (Mb), 161
Myrosinase, 184

N
Naphthoquinones, 143
Naringin, 88
Naringinase, 88
Native starch, 36, 38
Natural flavours, 175
Niacin, 146
Non-competitive inhibition, 81
Nucleation, 111
Nucleus, 111

O
Oil, 97
Oligomer, 66
Oligosaccharide, 32
Omega fatty acids, 99
Osmosis, 6
Oxymyoglobin, 161

P

Pantothenic acid, 146
Pasting curve, 37
Pectin, 42
Pectinase, 85
Pectin methylesterase, 85
Peptide bond, 60
Peroxidases, 88
Peroxyl radicals (ROO·), 105
Pheophorbide, 163
Pheophytin, 163
Phospholipids, 102
Photo-oxidation, 105
Phytic acid, 150
Phytochemicals, 150
Phytol, 162
Phytosterols, 104
Pigment, 160
Polygalacturonase, 85
Polymorphic transition, 114
Polymorphism, 112
Polypeptide, 60
Polyphenol oxidase (PPO), 124
Polysaccharide, 32, 33
Polyunsaturated, 98
Porphyrin ring, 162
Pregelatinisation, 38
Primary structure, 63
Prolamin, 38, 62
Prosthetic group, 76
Proteases, 86
Protein, 61
Pullulanase, 84
Pyranose, 23
Pyridoxine, 146

Q

Quaternary structure, 66

R

Recrystallisation, 112
Reducing sugars, 28
Reduction, 28
Rennin, 86
Resistant starch (RS), 38
Retinol, 142
Retrogradation, 37
Rhamnogalacturonan-I (RG-I), 42
Riboflavin, 145

S

Salting in, 70
Salting out, 70
Salty, 178
Saturated, 98
Schiff base, 60
Secondary structure, 63
Selectivity, 77
Shortening, 115
Smooth region, 42
Solid fat content, 116
Sour, 178
Specificity, 77
Sphingolipids, 103
Sphingomyelin, 103
Sphingosine, 103
Stabiliser, 33
Stanols, 104
Starch, 34
Starch granule, 34
Starch paste, 37
Stereoisomerism, 20
Stereospecific numbering, 101
Steric, 24
Sterols, 104
Steviol, 29
Storage proteins, 62
Strecker aldehydes, 132
Strecker degradation, 132
Substrate, 74
Sucrose, 31
Superimposable, 20
Surimi, 88
Swelling, 36

T

Tautomerism, 26
Tautomers, 25
Tempering, 115
Terpenes, 184
Terpenoids, 184
Tertiary structure, 65
Thiamine, 145
Tocopherols, 143
Tocotrienols, 143
Trans, 99
Transglutaminase, 88
Triacylglycerol (TAGs), 101
Trigeminal nerve, 180
Tuning fork, 112

U
Umami, 180
Uronic, 28

W
Water activity, 7
Waxy maize, 36

X
Xanthan, 45
Xanthophylls, 163

Z
Zwitterion, 59

Printed in the United States
by Baker & Taylor Publisher Services